Unidentified Aerial Phenomena

Unidentified Aerial Phenomena by Shane Hurd

© 2020 Shane Hurd. All rights reserved.

This book or any portion thereof may not be reproduced or used in any manner whatsoever without the express written permission of the publisher/author. except in the case of brief quotations in critical articles or reviews.

Although every precaution has been taken to verify the accuracy of the information contained herein, the author/publisher assume no responsibility for any errors or omissions. No liability is assumed for damages that may result from the use of information contained within.

Unidentified Aerial Phenomena: A Beginner's Guide to Researching UFOs / Shane Hurd — 1st ed.

ISBN 9798643273004
1. UFOs 2. UAPs 3. Society 4. Science

Cover by Jason McClellan
Printed in the United States of America

First published in the United States of America by Rogue Planet.

http://RoguePlanet.tv

Unidentified Aerial Phenomena

A Beginner's Guide to Researching UFOs

By

Shane Hurd

Acknowledgements

This book is dedicated to my dear wife Donna, daughter Whitney, and granddaughters Breelyn and McKinley who have supported me throughout this crazy adventure. I also must thank my dear friend and mentor Jason McClellan who has kept my feet on the ground, while at the same time inspired me to appreciate all things strange. Finally, I must also thank my MUFON family, including Jim Mann, Stacey Wright, and Dennis Freyermuth who have given me immeasurable opportunity. Thank you, thank you, thank you!

Contents

Foreword 1
Introduction 5
What is UAP? 13
Misidentification – A Common Occurrence 23
Hoaxes – A Small Percentage, A Big Problem 33
The Scientific Method 39
Forms of Evidence 47
Classic Cases 57
Ufology: Separating the Wheat from the Chaff 77
The Cover-up is Real 85
Political Science and Ufology 93
Follow the Evidence 119
Sources 127

Foreword

By Jason McClellan

UFOs are endlessly fascinating. Mysterious objects in the sky have been documented since the beginning of written language, and continue to be observed by people from all walks of life, all around the world, every day. Although most UFOs are eventually identified, there is still a sizable percentage that remain unexplainable. Some of these objects display characteristics that defy logic and our understanding of physics. Fighter pilots have attempted, but have been unable, to intercept mysterious aerial objects. Governments around the world have attempted to solve the UFO mystery. But answers to this perplexing enigma are elusive.

I'm an incredibly curious person. I'm fascinated by the unknown and by mysteries of the universe. When I first realized that unidentifiable objects are routinely present in our airspace, that the military has tried and failed to intercept these things, and that supposedly nobody knows what these objects are, my fascination by UFOs was piqued.

I've been researching and writing about UFOs for more than a decade. But, when I started my journey down that rabbit hole—my quest for information/knowledge related to the subject of UFOs—I was pretty lost. The UFO topic is complex, extending into a wide range of subtopics (e.g., simple UFO sightings, alleged alien contact, crop circles, cattle mutilation). There are countless books, TV shows, documentaries, podcasts, and websites with information about UFOs. And, there is an endless supply of lecturers, authors, and researchers who are more than eager to provide you with all the answers to the UFO mystery . . . answers they've come up with that conform to their personal beliefs. These issues made my entry into UFO research extremely daunting and horribly confusing.

Those same challenges I faced when I started down the

UFO road more than a decade ago still exist today, and, in fact, are probably even greater challenges for people like you starting down that road today. Fortunately, you have a valuable resource that I didn't have—this book. Having an introductory guide like this book would have made my entry into UFO research so much easier.

Shane Hurd is a veteran UFO researcher who approaches the UFO topic with an objective, scientific mindset. Unlike most people who research the UFO topic, Shane is also an active investigator who goes out into the field and physically investigates UFO cases. His knowledge and his experiences in this field are invaluable. And, as this book demonstrates, Shane is always eager to share his knowledge with other researchers. This book is an beneficial resource for anyone interested in researching the UFO subject. It provides you with a quick, honest introduction to the world of UFO research, it offers suggested investigation procedures, and, perhaps most importantly, it highlights a few of the obstacles and potential problems that UFO researchers encounter, and offers suggestions for how to avoid, or deal with, these issues.

Good luck on your journey into the world of the unexplained. You have this book, so you already have a head start, and are on much better ground that I was when I began my exploration.

Introduction

I was born in 1961. And, like just about every other kid in America, I wanted to be an astronaut. It was an exciting period of American history. The space race was in full swing. The American public was all in, supportive of the country's space ambitions. Scientific and technological achievements were multiplying in leaps and bounds. Schools took advantage of the interest and excitement and added science, computer, rocketry and other space programs to their curriculums. I will never forget the look of the mighty, gleaming white and black Saturn 5 rocket towering over the landscape against the bright blue Florida skies, shaking thunderously, spewing white smoke and fire! Truly inspiring.

In 1970, the United States population was about 205 million. It is estimated that the Apollo space program employed nearly 500,000 people. Mathematically, that means one in every 400 people worked on the space program! If you considered an average family of four people, that means one in 100 families had a direct connection to the space effort. It is no wonder that interest in space and technology was in the consciousness of nearly the entire country.

This interest was also reflected in other facets of American culture. Pop culture was greatly influenced by the interest in all things space. For example, fashion, art, automotive design, toys, music, movies, books, and comic books reveled in the space theme. And the list could go on. I remember as a young boy having a fascination with G.I. Joes and Major Matt Mason toys. These toy series included the astronaut figures themselves and a whole line of cool accessories such as space capsules, lunar rovers, moon bases, and space suits.

Who could forget songs like "Space Oddity" featuring Major Tom in 1969 by David Bowie? Or Elton John's "Rocket Man"?

These songs were like a soundtrack to the space generation. Other prominent displays of the interest in space were the movies of the 1960s and 1970s. The now classic *2001: A Space Odyssey* set the standard for space exploration and technology films. The 1950s, 1960s, and 1970s exploded with science fiction movies largely based on space themes. Of course, the genre also included imaginative themes of intelligent life elsewhere and the potential interactions and outcomes we may experience as a result of our space exploration. This gave birth to science fiction movies such as *Close Encounters of the Third Kind*, *E.T.: The Extra-terrestrial*, *Men in Black*, *Star Trek*, and the *Star Wars* franchise. By the time the space program wound down, people were beginning to explore these other themes in earnest. Yes, alien life and UFOs were also establishing themselves in the consciousness of the country as well.

Of course, the UFO phenomenon was nothing new. Alien life was the stuff of fiction from the likes of Jules Verne and H. G. Wells since the late 1800s. But, during and after World War II, the sightings of some very real, but unexplainable, objects in the sky brought them off the pages of books into our reality. Many are acquainted with the reports of encounters by military pilots of "Foo Fighters" from both the Pacific and European theaters of World War II. These bright-light orbs would fly maneuvers around planes in flight in a manner that suggested intelligent control. Yet, they seemed to be observing rather than interfering with the planes. So, the military did little more than take reports, and, to our knowledge, has never identified them.

Then, more sightings came at home in America. In the early 1940s, Cape Girardeau, Mississippi and Los Angeles, California, preceded a string of sightings largely considered to be the start of the modern-day phenomenon beginning with the Maury Island incident and Kenneth Arnold's famous sighting over the Cascade Mountains of Washington State. What is arguably the most famous sighting—Roswell, New Mexico—followed. From there, the list goes on and on of very public and well-documented cases that continue to this day.

As a result, UFO-related phenomena are entrenched in our society. It has seen peaks and valleys in its popularity, but it is a part of the fabric of our culture. I don't seem to recall the people in my life at the time to be overly skeptical or overly accepting of the phenomenon, but maybe because I was just a kid. I do recall thinking UFOs a possibility.

Of course, once I grew into adulthood, experienced higher education, the workplace, and exposed to more thoughts, ideas, and opinions of others, I realized there was indeed skepticism on the topic. It largely fell off my radar as a serious topic until March 13, 1997.

You may recognize this date as the evening of the "Phoenix Lights" incident. I was a resident of Phoenix, Arizona and even know what I was doing that night. Our friends who lived about 40 miles north of the city invited us to dinner and to spend the evening looking at Halley's Comet in their telescope. It was very dark that far from the halo of light pollution from the city and we had a great view. Ironically, we saw neither the objects flying over the city, nor the light display over the Estrella Mountains some 60 miles away. However, the incident sparked a media frenzy for months from both local and national news, and it was in my own back yard! Well, this renewed my interest in the topic and I began to keep my eye open for information about that sighting and others.

The internet was still in its relative infancy, and I wasn't intimately involved with it either recreationally or for business. So my source of information was largely television. But, not surprisingly, there began to be many programs related to UFOs. So, I did watch what I could casually as I went about the business of marriage, fatherhood, making a living, and life in general.

This did keep my interest alive until I had a little more control over my life and my time as my child grew and my career stabilized. In 2010, I happened across a book by Bill Birnes and Phillip Corso titled, *The Day after Roswell*. I was curious and was already developing an appetite for more information than the casual UFO program such as *UFO Files*. I also read, *UFOs: Generals, Pilots, and Government Officials Go on the Record* by Leslie Kean. I highly recommend you read this book. It is among the most professional and cogent discussions of the topic. I followed that book up with Richard Dolan's powerful two volumes of *UFOs and the National Security State*. Once I read these books, I came to my first solid conclusion of the topic: There is a wide variety of information available—some of it dubious, some of it plausible, and some of it fascinatingly factual. But, more than that, I was convinced of the reality of the phenomenon. Something unexplainable is happening before our very eyes!

Thus, my journey began. I already felt there was at least something occurring that merited my interest, but I was not going to waste my time with fantasy. I have spent the better part of forty years in the construction, civil engineering, and information technology fields. I work for a local government as a civil employee. I wanted facts, I wanted data, and I wanted proof! As a manager, I also knew that the only way I was going to get it was dig in and do the research myself.

However, the search for facts, data, and proof has not been easy. Believe it or not, there is an overwhelming amount of information available on unidentified aerial phenomena. As I alluded to earlier, some of the information ranges from outright crazy talk, to very professionally researched and scientifically-presented facts. Of course, there is a lot of information that is somewhere in the middle. So, it is up to you to discern fact from fiction.

How do you do that? How do you calibrate your BS meter? How do you research this topic? Well, like any form of research, you must multi-source the information. You must also verify the credibility of the source of the information. One way you can do this is to compare your source of information with known credible sources. Think of it as a form of peer review. You will find that there are UFO researchers who police the field of research. The reason this is done is to protect the integrity of the topic and the legitimate research community.

Another thing to give thought to as you begin researching the topic is to consider the media format in which the information is being presented. Some forms of media are more likely to be misused than others. For example, like everything else in this world, money makes it go around. The study of the UAP phenomenon is no exception. Many people today are not concerned with the truth of the phenomenon. It is simply a way to make money because they can easily exploit the interest of the public.

The most effective media format to do this is the internet. There are many YouTube channels, websites, social media pages, and other media that are clearly purveying sensationalistic falsehoods in the name of real UAP. With today's video and photo software, it is cheap and easy to create very convincing images. So, be careful, use good judgement, if it looks too good to be true, then it is likely a hoax.

Books are a two-edged sword when it comes to credibility. I have found that most books are sincere efforts to promote a genuine understanding of the UAP research. Publishing a book is not cheap, nor easy. It requires a great deal of research and expense, which is the product of someone's true interest in the topic. There is no guarantee that a book is more factual than a video or some other presentation. But odds are that the book came as a result of a great deal of effort, whereas a video on the internet can be produced faster, cheaper, reaches more people, and generates a lot more revenue.

Among my most favorite UAP related media is government reports and documentation. The authenticity can usually be proven. They are generated based on a real issue and the governmental organization's response to it. Now, the interpretation of the purpose of these documents can be disputed, but I place a lot of credibility on the provenance of this kind of evidence.

Finally, the question you will likely be asked is, "Do you believe in UFOs?" This is an absurd question, because UFO means "Unidentified Flying Object." It is not a question of belief that people see flying objects they cannot identify. It is a fact of life. People have misapplied the term "UFO" to mean something identified—namely an alien craft. One must understand that the real question being posed is, "Do you believe that we are being visited by alien life?"

At this point in my research and understanding of the data, my view is there is an abundance of evidence indicating there are machines of unknown origin flying in the skies and low orbit of the Earth. However, this topic is complex. There are tangent topics often conflated with UAP. There is the paranormal, the abduction phenomenon, multi-dimensional or multi-universal theory, ancient astronaut theory, and a host of other so-called related topics. So where does one begin?

As any journalism student is taught, an investigation for a story must start with answering the five "W" questions: what, where, when, who, and why. I think we must start with the "what," "where," and "when" questions since they can be answered by the data, by observation, by the scientific process. That is where I recommend your research journey begin. If those questions are answered definitively, then the answer to the "who" and the "why" questions will become self-evident.

So, are you are ready to begin your search for the facts, the data, and the truth to answer for yourself the question: Unidentified Aerial Phenomena – What is it?

Note to Self
How do I calibrate my BS meter? Ask the five "W" questions: what, where, when, who, and why.

Chapter 1

What is UAP?

Unidentified Aerial Phenomena (UAP) is the new term used to describe what was formerly referred to as Unidentified Flying Objects (UFO). Why the change? The United States Air Force coined the term "Unidentified Flying Object" in 1953. Its initial definition was "any airborne object which by performance, aerodynamic characteristics, or unusual features, does not conform to any presently known aircraft or missile type, or which cannot be positively identified as a familiar object." Simply put, an object in the sky that the viewer could not identify. This is a clinical definition that served to allow terminology that could be used to properly characterized reports made by the public.

However, over the years, that definition has been lost and a new unscientific context was created. The term UFO soon became synonymous with "flying saucer" or "alien craft," which would ironically be "identified." This misnomer contributed to the topic becoming considered unscientific and ridiculed, not only in scientific, government, and military circles, but by the general public as well.

It appears the term UAP was used by the United Kingdom's Ministry of Defence in the 1990s. UAP was frequently used by the British defense intelligence staff while their civilian colleagues continued to use UFO. However, UAP was not new, because the phrase "aerial phenomena" has been in use by the RAF since at least 1952. Nick Pope was assigned to the UFO desk as an analyst from 1990 through 1994. In a memo written to enlist the support of MoD leadership to further case research, he used this term instead of the term UFO. It was his feeling that the term UFO would not have been taken seriously and would have resulted in a denial of his request. Well, the term UAP worked. He received approval and the project was approved.

The acronym UAP is used throughout the study commissioned by the MoD entitled "Project Condign." The creation of a less controversial term—UAP (Unidentified Aerial Phenomena)—to describe inexplicable incidents, was the solution to the consequences of using the loaded term "UFO." While UAP appears in official documents as early as 1962, soon ufologists began to adopt a version of the term. UAP, with the meaning "unidentified atmospheric phenomenon," was used by UFO investigators Jenny Randles and J. Allen Hynek during the late 1970s. According to *UFO Study: A Handbook for Enthusiasts* by Jenny Randles in 1981, she stated that they felt that UAP was a better term to use to interest scientists because "it presumed less and was more accurately descriptive than UFO, which, both by its use of the word object and by years of presumed application now inferred a material craft, usually a spacecraft, in many people's minds."

So, there began to be consensus that the term UAP should be used instead of UFO by both official and unofficial organizations and researchers. I highly recommend that you use this terminology as well. It carries much less stigma and really is a more accurate description of what is being seen.

The term UAP has also been used by the National Aviation Reporting Center on Anomalous Phenomena (NARCAP) since 2000. Led by Executive Director Ted Roe, NARCAP is an aviation centric organization interested in providing a safe reporting platform for private, commercial, and even military pilots who have seen anomalous phenomena in their flight activity. It is especially concerned about flight safety. If pilots are ridiculed or their credibility and career are threatened by making such reports, the potentially dangerous conditions go unreported and may therefore pose a threat to safe flight operations. In addition, this organization is also interested in collecting data and providing a scientific-based process to the identification of these phenomena.

According to its website, www.narcap.org, the organization defines, "An Unidentified Aerial Phenomena, UAP, is the visual stimulus that provokes a sighting report of an object or light seen in the sky, the appearance and/or flight dynamics of which do not suggest a logical, conventional flying object which remains unidentified after close scrutiny of all available evidence by persons who are technically capable of making both a technical identification as well as a common sense identification, if possible."

The Chief Scientist for NARCAP is Dr. Richard F. Haines, who holds a PhD in Psychology and served as a NASA research scientist appointed Chief of the Space Human Factors Office (1986-1988). He directed the development of EVA spacesuits, habitability design for Space Station Freedom, and spacecraft window design, per his biography on narcap.org. But his interest in UFOs spans decades and led him to create NARCAP. The organization also has an extensive research archive including international reports, technical reports, topical studies, and is associated with several international partners such as The Committee for the Study of Anomalous Aerial Phenomena (CEFAA) of Chile. This is an excellent source of credible data and one I highly recommend you make use of in your study of UAP.

Interestingly, John Podesta, a politician and advocate for the release of UFO data by the United States government has used this term extensively. In fact, he was the campaign chief for 2016 presidential candidate Hillary Clinton. And, for the first time, a candidate embraced the UFO topic. She had agreed to open what files she could on the UFO topic, if elected. Her campaign had commented on the topic numerous times, she has been interviewed by various media, including a stint on the *Jimmy Kimmel Show* where she made a point of correcting his use of the term UFO and indicated the new term is UAP!

It's hard to imagine that the term UFO is now being taken seriously when you consider what happened to presidential candidate Dennis Kucinich in a 2008 presidential debate where he was asked about a UFO sighting he had with actress Shirley

McClain, as mentioned in her 2008 book *Sage-ing by Age-ing*. The poor guy was virtually laughed off the stage when he tried to give a legitimate response! It is thought to be one of the things that killed his candidacy.

Another advocate of the term UAP is investigative journalist and author Leslie Kean. Leslie is the author of the book *UFOs: Generals, Pilots, and Government Officials Go on the Record* (I mentioned this book earlier in the introduction). She is also a board member of the recently created organization UFODATA.

Her biography on UFODATA.net lists her UAP related accomplishments as follows: "Leslie Kean is an investigative journalist and author of *UFOs: Generals, Pilots and Government Officials Go on the Record* (Crown Publishing Group, 2010), a New York Times bestseller which includes a foreword by John Podesta and has received endorsements from numerous scientists, such as former White House OSTP Director Neal Lane and physicist Michio Kaku."

Since the release of *UFOs: Generals, Pilots, and Government Officials Go on the Record*, Kean has been featured on news programs such as CNN, MSNBC, FOX, The Colbert Report, NPR; and in print media such as USA Today, US News and World Report, the Columbia Journalism Review, the Huffington Post, and The Arizona Republic.

In 2011, she co-produced a feature documentary for the History Channel based on her book. Since then, she has contributed a series of stories to the Huffington Post. Previously, Kean worked as a radio producer/host on Pacifica Radio and published articles in the Boston Globe, the Journal for Scientific Exploration, International Herald Tribune, Sydney Morning Herald, Globe and Mail, and The Nation, among many other publications. I also recommend ufodata.net as a credible resource for your study of UAP. This recently formed organization is a collection of very credible scientists, engineers, PhDs in psychology, astronomy, astrophysics, political science, and government officials including defense and intelligence officials from the United States. In addition to the credibility of the organization's leadership, the website also states the purpose and methodology of the research to be conducted:

"As with any real phenomenon, UFO/UAP (Unexplained Aerial Phenomena, which we use interchangeably with UFO) - whatever they might be - can be studied scientifically so long as we have the proper equipment to measure their physical characteristics." Evidence such as radiation and ionization are areas of scientific scrutiny. Spectroscopy is the study of radiation intensity as a function of wavelength. A rainbow after a rainstorm is an example of natural spectroscopy, with the Sun's visible light split up into its component wavelengths.

For the UFODATA group, the primary study of a UAP spectrum will be in visible light, captured with cameras and an attached spectrograph.

But there are other physical characteristics that are potentially important as well, such as a magnetic field associated with a UAP. Magnetic fields can be generated by either moving electric charges or by elementary particles that have a synchronized spin in a material such as a magnet). A sufficiently strong magnetic field requires a large energy source and detecting a magnetic field during a UFO event would allow the estimation of the source energy."

As a matter of reference for your research, the UFODATA folks make the point that all energy comes in the form of particles, and is carried by photons. Other forms of energy are associated with various types of elementary particles, including alpha and beta radiation. An alpha particle consists of two protons and two neutrons, bound together. These particles are emitted through radioactive decay, among other potential mechanisms. When it is said that something is "radioactive," it is emitting these particles, which can at high enough intensity be quite dangerous such as near a nuclear reactor.

It is generally not thought that being near a UFO is dangerous, yet there have been reports of physical effects on human beings from close proximity to some UFOs. It is useful to measure such radiation, which can be done with a Geiger counter.

Other potential effects emitted from a UFO can include a

static charge to build up on surfaces. This type of charge is like a minor shock after walking across a rug and then touching a conductor, and you get zapped! Not deadly per se, but it can hurt! Of course if there is a significant charge, such as is demonstrated with lightning, it could be fatal!

MUFON Investigators always collect geographic and weather-related data because temperature, relative humidity, atmospheric pressure, and wind speed and direction all can have an important role in establishing the possible explanation for a UFO sighting.

The UFODATA scientists also note: "Reports of UFOs have often indicated that their appearance, including the intensity of their visible light, along with their color, often changes, sometimes quite rapidly. This variability is demonstrated by the fact that UFOs have been reported to pulsate, with their emitted light going on and off; however, that doesn't mean that the phenomenon itself has really disappeared. All the other wavelength ranges, and these other physical properties, must be measured simultaneously to see how they change as the UFO's appearance varies. So, we will measure the properties of the UFO continuously when the station is 'triggered' into action."

The method they intend to use to collect the data to hopefully one day identify UAP is the use of instrument suites compiled into a "Station." It is further described in this excerpt from the Technology page as follows:

"UFODATA stations will consist of a suite of instruments to measure many physical characteristics, of both the UFO/UAP (Unexplained Aerial Phenomena, which we use interchangeably with UFO), such as its visible light and spectrum, as well as environmental measures, such as local magnetic field, or ambient temperature and air pressure.

The prototype station will have a basic set of instruments, which will be extended once the technical, software, data storage, and communication details have been worked out and resolved, along with testing of this first set of instruments.

Stations will also have computing hardware, data storage

hardware, internet connectivity, and will run various types of control and other software to manage the instruments and their integration. They will also have a dedicated power source, which will initially be a local connection to the power grid. Eventually, we expect stations to be powered by batteries charged by solar power so that they can be located in relatively remote regions. Fortunately, UFOs have been seen everywhere there are people, and so the stations don't have to be placed on top of isolated mountains, or in uninhabited desert or forests, to maximize the chance of detecting and recording data about the UFO phenomenon."

As you can see, this organization is going to provide hard data on anomalous events. I have quoted extensively from them because they represent what ufology has been striving for all along. This is an absolutely exciting development in ufology, and I highly recommend you include this in your research. At the time of this writing several other groups are in development of such UFO detection instrument suites.

When you think about it, UAP has a broader definition that describes what is being seen in our skies and space than the term UFO (flying saucer). For example, a common sighting is "lights in the sky." These lights not only radiate in an unusual way, but also move in an unusual way. However, one cannot say they represent a defined "object." There are other unusual phenomena such as spirals in the sky. Again, these are not distinguishable as an object, but certainly qualify as Unidentified Aerial Phenomena.

Having a good understanding of the terminology and the political history behind it can help you in your research. This topic is very much about history, context, and other political influences. So, get to know these contexts and how they influence the information you are researching. Clearly, as the information above indicates, many credible and serious people and organizations are expending a great deal of effort on the research of UAP. It is serious business.

Now, I must ask you: Have you personally seen something in the sky you could not identify? Is this the basis for your interest in UAP? If so, you are not alone.

Particularly since the Kenneth Arnold sighting of several boomerang-shaped craft over the Cascade Mountains in June of 1947, reports have been made of unidentified aerial phenomena across the world. Although there are recorded sightings in newspapers in the 1800s across the United States, the frequency and intensity of sightings rapidly increased in our modern age. In fact, there have been tens of thousands, if not hundreds of thousands of reports since that time. There is no doubt that something is going on. What is it?

Many people and organizations view Unidentified Aerial Phenomena as a reality. The United States Air Force, Central Intelligence Agency, Federal Bureau of Investigation, National Security Agency, and other organizations—public and private—have had programs to analyze the many reports from very credible individuals from all walks of life. The UK, Chile, France, and others have projects, programs, and reports reflecting that many other nations are experiencing the same thing. Official investigations such as the USAF's Project Sign, Project Grudge, Project Blue Book, the University of Colorado Condon Report, and France's COMETA Report contains a significant percentage of unexplainable cases. These unexplained cases are not due to a lack of evidence. On the contrary, the evidence is sufficient to determine that the sighting was truly unexplainable by conventional means!

For example, of the 12,618 cases filed under the United States Air Force's Project Blue Book, 701 cases were unexplained. If even one of those proved to be identified as from intelligent life not from earth, our entire universal view would change.

Every year, tens of thousands of credible people are seeing something in the sky that they cannot explain. What is it?

Note to Self
UAP is the new modern term for UFO.

Chapter 2

Misidentification – A Common Occurrence

Ok, I really hate to do this to you, but it is necessary, and we might as well get it out of the way now. It's back to school—Science 101 to be exact! You are obviously interested enough in the study of UAP that you are investing time, thought, effort, and money into the subject, so you better do it right.

Although I am devoting an entire chapter later in the book, I feel I must at least mention the *scientific method* of approach to your research of the UAP topic now, especially as it relates to the issue of misidentification.

That means you must employ the *scientific method* to your study of UAP. In addition, you must know the method to help identify that the sources of information you study have also employed the scientific method in their research.

If so, then you can have a high degree of certainty about the quality of the data and conclusions you and they have made. Otherwise, this is just of casual interest to you and you are ripe for the picking by hoaxers and debunkers. You must really "know your stuff" so you can defend yourself. So, let's take a little time to summarize the "Scientific Method" and how it can be applied to the misidentification of UAP.

The scientific method is the process in which science is performed. It can build on previous knowledge and develop a more sophisticated understanding of its topics of study over time. The overall process involves making hypotheses, deriving predictions from them as logical consequences, and then carrying out experiments based on those predictions to determine whether the original prediction was correct.

First, we figured out fire, then lightning, then electricity, then nuclear energy. A process of building on prior knowledge and understanding. So too with the study of UAP. Don't reinvent the wheel. There is a great deal of valuable data and knowledge amassed from past cases, studies, projects, and reports. Make good use of this data in your own research of UAP.

Scientific inquiry generally tries to obtain knowledge in the form of testable experiments that can be used to predict the results of future experiments. This allows scientists to gain a better understanding of the topic being studied, and to use that understanding to intervene in its causes. The better the explanation is at making predictions, the more useful it is, and the more likely it is to continue explaining a body of evidence better than its alternatives. The most successful explanations, which explain and make accurate predictions in a wide range of circumstances, are the scientific theories.

But this is the rub with the study of UAP. It is very difficult to even observe UAP. It happens randomly in a time and place of its choosing, not ours. So how can you possibly replicate an observation, let alone test its cause? This is the very reason that programs such as the UFODATA project mentioned in Chapter One are so important. This project seeks to assemble instruments and deploy them in the field to record *provable data* on UAP activity. We then may be able to record an observation and its physical data in real time for use in quantitative and qualitative analysis. There is physical data associated with UAP events.

Many people are surprised to learn that there have been residual physical effects with many UAP cases. Evidence such as broken tree branches, burned plants, disturbed soil, radiation, chemical traces, fires, smoke, slag, photographs and motion pictures on analog film, radar system recordings of radar returns, electrical system disruptions, and more. People and animals have been injured and killed by contact with UAP. So, there is a great deal of good data and evidence to work with, but it still is a challenge.

So, as the title of this chapter suggests, misidentification is a possible explanation for some UAP sightings. This has been proven true through the scientific method. Even though Project Blue Book conducted by the United States Air force from 1952 to 1969 was not conducted in a formal scientific method, it did accurately identify some UAP sightings. It is worth noting that they were mostly concerned with dismissing the phenomenon to settle down the public interest so as not to incite panic or acknowledge the USAF had no control over the phenomenon.

Allegedly, the substantive cases were kept out of Project Blue Book and investigated separately. Of the 12,618 cases in the report, all but 701 were identified as prosaic causes such as weather balloons, conventional aircraft, the planet Venus, lenticular clouds, and other natural phenomena. So, let's briefly familiarize ourselves with a few of these natural and man-made phenomena that are often misidentified as UAP.

Optical illusions are a fact of life. The combined human eye and brain are an instrument that can be easily fooled. Unfortunately, sight is one of the least reliable senses we possess. An optical illusion is caused by the eye and characterized by visually perceived images that differ from objective reality. The information gathered by the eye is processed in the brain to give a perception that does not match a physical measurement of the stimulus source.

Optical phenomena include those arising from the optical properties of the atmosphere, of nature, of objects (whether natural or human-made), the optics of our eyes, or the misinterpretation of what we see by the brain.

There are many phenomena that result from either the particle or the wave nature of light. Some are quite subtle and observable only by precise measurement using scientific instruments. One famous observation is the bending of light from a distant star by the gravity of the Sun observed during a solar eclipse. This demonstrates that space-time is curved, as the theory of relativity predicts.

There are three main types of optical phenomena: *literal* optical illusions that create images that are different from the objects that make them, *physiological* illusions that are the effects of excessive stimulation of a specific type (brightness, color, size, position, tilt, movement), and *cognitive* illusions, the result of unconscious inferences.

One example of a trick one's eyes can play upon us is what is termed an "afterimage." Have you ever looked at the blinds hanging in the window with the light peeking through the slats, only to look away and still see the horizontal bands of light in your field of view? That's an afterimage.

Another example is a "halo." Have you ever been driving at night only to see the headlights of oncoming traffic as a bright light in the center, but with a bright ring around it? Or the same scenario but with trails of light? It's an optical illusion known as "light streaking." How about "Parallax." This is seen when you are driving down the road and it appears the fence posts, the trees, and mountains are all moving in relation to each other as you move down the road.

"Refraction" is the illusion you get when a straight object like a fishing pole is dipped in the water and appears to be bent where it enters the water. This is due to the density of air versus water and produces the effect. Or the effect when the moon looks larger at sunrise and sunset than when it is high in the sky.

"Auto-kinesis" is another form of optical illusion from a cognitive standpoint. Have you ever been focused on a star and it began to wiggle or otherwise move in some way? This appears to be phenomena related to visual perception and the brain trying to reconcile motion as it relates to fixed objects. It is part of the process of keeping our balance while moving. Of course, it makes it possible to misinterpret the stationary star as a moving object, or UAP.

So, it is a fact that some optical illusions could be responsible for some UAP sightings. However, the devil is in the details. Literal, physiological, or even cognitive optical illusions do not make a triangular set of indentations in the soil, elevated radiation levels, broken tree branches, radar images, saucer shapes or triangular objects floating a few hundred feet above one's head shooting beams of light on the ground at one's feet.

I think you get the point. One must use logic, reason, and common sense when evaluating a UAP case. The details matter! And no, optical illusions are not the explanation for all UAP sightings.

Natural Phenomena have certainly been misidentified as UAP. In many cases, the more prosaic explanations of the "planet Venus," the "moon," "lenticular clouds," and other astronomical or meteorological phenomena have been correctly identified as the cause for some reported UAP sightings. Some of these are either rarely seen, never seen, or are seen completely out of their normal context, which can be disorienting and result in a misidentification.

Plasma is a natural phenomenon that could explain some UAP. One form of plasma is called "ball lightning." Little is really understood about this atmospheric, electrical, plasma related phenomena. Attempts at recreating the effect in the lab have only produced effects visually similar, but not proven to be related to the natural phenomena. There have been many reports over the centuries of these balls of light appearing from the size of a pea to several meters in diameter, usually associated with a thunderstorm. Perhaps this accounts for some UAP sightings, perhaps not.

Another form of plasma is "St. Elmo's fire." This is a form of plasma that is sometimes generated from aircraft wings passing through an atmosphere of electrically charged ions resulting in a blue or violet glow along the wings edge.

Some feel these may be an explanation for the so called "Foo Fighters" of World War II. However, while this may explain some of the reports of lights in the sky or floating orbs, it certainly does not accurately describe or account for all UAP sightings.

Lightning storms known as Sprites could account for some UAP reports. These are the jellyfish looking lightning storms that appear very high in the atmosphere. In fact, they were first recorded on film accidently in 1989 by scientists. Some range from 35 to 80 miles in altitude, which is much higher than the normal 7 to 10 miles where lightening occurs. Again, with no experience or context, these can appear very strange and may account for some UAP sightings.

Of course, there are other meteorological phenomena such as halo's, lenticular clouds, hole-punch clouds, mirages, and temperature inversions. Or as already discussed, astronomical objects such as the planet Venus, the Moon, meteors, comets, or constellations have been misinterpreted as UAP.

It would be quite arrogant for us to think we know all there is to know about physics, the natural world around us here on earth and of the universe. So, some known or unknown natural phenomena may account for some UAP sightings, but not all.

It is important to note that these explanations have also been intentionally and falsely misused to debunk a sighting of UAP, often by authorities such as the USAF.

The intent of Project Blue Book by the United States Air Force from 1952 until its closure in 1969 was to explain away UFO reports for fear that the public would panic. Little investigation was performed. The majority of those explanations were misidentification of natural or manmade objects. But, this ploy was soon realized by the public and they started to get fed up with the intelligence insulting explanations.

One particularly famous and much maligned example was the "swamp gas" explanation offered by Dr. J. Allen Hynek of Project Blue Book in 1966 as part of a UAP wave in Michigan. This obfuscation so incensed the people of Michigan, who new darn good and well that objects were seen by hundreds of people, that congressional hearings were initiated by then Michigan Congressman and future president Gerald Ford. This was largely because of the outcry of the people of Michigan who finally had it with the Air Forces obvious and insulting debunking!

Aircraft have been misidentified as UAP. There is little doubt that airplanes, helicopters, balloons, missiles and other man-made objects account for many sightings. Weather balloons can look quite strange because of their shape, reflectivity and altitude in the sky.

I recently had my own experience witnessing a weather balloon that had been released from White Sands AFB in New Mexico. It was about 6:30 p.m. and I was out feeding the horses when I looked up to the east and saw this brilliant shiny orb high in the sky. It was very reflective because the sun was setting in the west and shining directly on the balloon. At first, the balloon appeared as a round orb. However, as it got closer I got out my binoculars and I could see that it was more tear drop shaped and had a payload hanging from it. I must say, at first it was a curious sight and it took me a few minutes and a pair of binoculars to figure it out. In fact, the nightly local news in Phoenix profiled it on the newscast because many people were calling to report it. By the time of the late news hour, the news stations correctly identified it as a weather balloon released from the base.

When an aircraft such as the U-2, the A-12, or the SR-71 would fly at an altitude of 80,000 feet, it appeared very strange. It was higher in the sky than airplanes normally flew in that day and age, so it was an uncommon sight and out of context. In some cases, the sun reflected on the plane even though the observer was standing in the darkness due to the curvature of the earth. This created an unexplained light moving in the sky. Again, this was very strange and unknown behavior for an aircraft.

There has been a long history of the development of "Black Project" aircraft performed in secret for security reasons. These include the U-2 high altitude reconnaissance spy plane, the A-12 Oxcart, the SR 71 Blackbird, F-117A stealth ground attacker, the B-2 stealth bomber, Tacit Blue, RQ-3 Dark Star, and many others. Today, with the development and deployment of Unmanned Aerial Vehicles in the military and even private sector, we could expect an increase in such misinterpreted sightings. Development and testing flights of these aircraft in the day and night skies no doubt has been reported as UAP. Yet, its worthy of note that these aircraft still operate on the principals of thrust and lift, have wings, tails, lifting bodies, rotors, propellers or any combination thereof, and the usual characteristics of known human flying technologies. However, not all UAP sightings can be explained by these misidentified aircraft.

It is also possible that some of these UAP are black project craft that demonstrate flight characteristics and propulsion systems not yet revealed to the public. These could be airframes in the form of disk shaped antigravity craft, triangular shaped antigravity craft, boomerang or delta shaped antigravity craft as reported in many UAP sightings. However, the size, performance, and the demonstrated use of this technology over populated areas of the country belies such a suggestion. Additionally, UAP sightings have been reported even before man was capable of flight.

Moreover, if a small, secret, elite segment of mankind possessed this technology and kept it secret from the rest of humanity, it would be a crime against humanity and the planet itself. This technology could be used as an energy source to free mankind of the perilous economic and environmental effects of the use of fossil fuels. There is a body of the UFO research community that believes there is evidence of a secret space program that has this technology and is using it as we speak. However unlikely it may be, it is possible this accounts for some UAP sightings.

It is also worth mentioning that there are thousands of objects in Earth orbit that can be seen with the naked eye, binoculars, and telescopes. These may be active satellites, the International Space Station, or simply space junk. Under the right conditions, it is possible that sightings of these objects could be misinterpreted as UAP. Of course, objects in earth orbit do not change directions, increase or decrease rapidly in altitude, or perform other maneuvers as reported in many UAP cases. So, not all UAP are explained by space junk, ice crystals, or meteors. I highly recommend you obtain a cell phone application that tracks satellites. You can immediately identify if something you are seeing in the night sky is a satellite, space junk or some astronomical object. Plus, they are just a lot of fun!

Interestingly, Project Blue Book Special Report Number 14 1, performed by the Battelle Institute and released in October of 1955, analyzed 3201 case study results from Project Blue Book and classified them according to the USAF investigation conclusions below:

Explanation	# of Cases	% of Total
Balloon	450	14.0
Astronomical	817	25.5
Aircraft	642	20.1
Miscellaneous	257	8.0
Psychological	48	1.5
Insufficient Info	298	9.3
UNKNOWN	689	21.5
Total	**3201**	**100**

Notice that "Insufficient Info" represented 298 cases or 9.3%. In other words, there was not enough data to make an identification. However, notice that is separate and distinct from the "UNKNOWN" 689 cases or 21.5% of the total cases. Debunkers like to conflate those percentages so it looks like 30.8% can't be identified due to insufficient information. However, 21.5% of cases did have enough data to conclude that the cause was conventionally *unexplainable and truly unknown*!

By definition, in this study by the Battelle Institute of the United States Air Force's Project "Blue Book," a whopping 21.5% of UFO sightings were confirmed UFOs! That's not a figure you hear the debunkers or dis-information agents touting.

So, that is 21.5 out of every 100 sightings cannot be explained by the 6 other categories representing 78.5% of the sightings. Put that in perspective in your mind. After every other reasonable, scientific, even far stretches of assigning an explanation to a case, more than 2 out of 10 sightings are genuine UFOs. If even one sighting was proven to be of an extraterrestrial source, our whole world would change. Isn't that reason enough to take this subject seriously? People have been tried and convicted for murder on less evidence than that!

So, let's consider another area in which you want to be educated as you research the UAP phenomenon. The despicable hoax!

> **Note to Self**
> Up to 21% of UAP sightings are truly unexplained.

Chapter 3

Hoaxes – A Small Percentage, A Big Problem

Hoaxes do explain some UAP sightings. Although, people seem to give this explanation a lot of credence. But, when you look at the statistical data regarding this explanation, it is plain to see that, when investigated, this accounts for a minimal amount of case study findings. Until recently, the only way to perpetrate a hoax was to lie or create a physical object, film it, and pass it off as a genuine object. That was difficult to do because, with just a little scientific scrutiny and common sense, the fraudulent nature of such lies or photos was usually exposed.

Now, more recently, with the invention and availability of computer software, an individual can create some convincing Computer-Generated Imagery (CGI) that, to the untrained eye, may appear valid. Although, with scientific scrutiny, these too can be identified as fraudulent. Sadly, today many people will see a photo or video on platforms such as YouTube and not take the time to research and determine the validity of the photo or video. Even when you do, it is also very difficult track down the source to verify the back-story. So, the internet is chock full of photos, videos, and websites that are complete fabrications. So, researcher beware!

Hoaxers do a disservice to all by creating a "boy who cried wolf" effect that causes all photo and video evidence to be doubted. This is a real shame because this can result in a loss of important scientific data and prevent serious scientists from studying the phenomena all together. It also provides a convenient, yet illogical "out" for debunkers to claim that "if one case is a hoax, they all must be hoaxes."

So, I have wondered why someone would perpetrate the hoax of a UAP event.

Let's start with Webster's definition of hoax for some insight. It defines a hoax as: *An act meant to trick or deceive; to cheat.*

I don't know anyone who would like to be tricked, deceived or cheated, do you? I don't think anyone would call that a good thing, or someone who would perpetrate it, a good person. There is nothing good about a hoax. In fact, some hoaxes can result in very dire consequences for those hoaxed or those hoaxing. People have lost a great deal because of hoaxes: Money, integrity, relationships, even their very lives. Frankly, perpetrators of hoaxes are bad people doing a bad thing.

What could be the motivation of such a harmful lie? The common thread for these perpetrators is a desire for personal gain. It may be for fame, fortune, or just recognition from peers. It may be of a desire to harm a certain person or group that they feel may have wronged them.

Let me say at this point that I make a distinction between a person who intentionally hoaxes others versus a true believer who is just gullible and unknowingly advances bad information in error. That is still a problem for ufology but is much less despicable on a moral level. So, just do your research and don't be that guy or gal!

I just have a hard time even imagining why someone would hoax. This moral deficiency fascinated me to the extent that I did a little research on the psychology of a person engaging in this and will share some of this information with you.

The reason I take this so seriously from the perspective of an amateur researcher of UFO phenomena is because I am searching for the truth. In a world of inherently ambiguous data, the last thing I need is some jerk trying to interfere with that search for truth by lying, deceiving, and cheating me out of it!

So, to help you know your enemy, here is what I have found out:

According to a 2008 article titled "The Fame Motive" in the *New York Times* by Ben Carey, this was said about a hoaxer's motivation:

"For most of its existence, the field of psychology has ignored fame as a primary motivator of human behavior: it was considered too shallow, too culturally variable, too often mingled with other motives to be taken seriously. But in recent years, a small number of social scientists have begun to study and think of fame in a different way, ranking it with other goals, measuring its psychological effects, characterizing its devoted seekers."

I think we are all becoming much more aware of the ability for people to achieve fame through reality television, YouTube, and social media. It is a significant part of our culture now.

Ben goes on to point out that "people with an overriding desire to be widely known to strangers are different from those who primarily covet wealth and influence. Their fame seeking behavior appears to be rooted in a desire for social acceptance, a longing for the existential reassurance promised by wide renown." This is one behavior you will most likely encounter within ufology by some proclaimed researchers, skeptics and social media observers and commenters. So, be aware and keep your own feet on the ground. Be motivated by curiosity, not fame.

Finally he cites a 1996 study, by Richard M. Ryan of the University of Rochester and Dr. Kasser, then at Rochester in which they conducted in-depth surveys of 100 adults, asking about their aspirations, guiding principles, and values, as well as administering standard measures of psychological well-being.

The interesting takeaways from the study include the participants in the study who focused on goals tied to others' approval, like fame, reported significantly higher levels of distress than those primarily interested in self-acceptance and friendship. Surveys done since then, in communities around the world suggest the same thing. Aiming for a target as elusive as fame, and so dependent on the judgement of others is psychologically dangerous.

So, hoaxers may or may not even be interested in UAP. However, the psychology of a hoaxer impels them to seek acceptance, recognition, and notoriety from a peer group through the attention they gain, positive or negative, through their hoax. Let's face it, ufology is a defined community and it would be easy to understand how to gain acceptance within it.

Of course, this psychology is not the only motive. We all know the almighty dollar makes the world go around, and the lure of quick, easy money is too tempting for some. As I mentioned, never before has there been as large and easy a platform to hoax and make money as in this day and age of the internet.

Don't get me wrong, I love the internet. As a MUFON Field Investigator, I do a great deal of research on the internet. It can be your friend. However, it can be your worst enemy. You can be fooled. You can give in to how easy it is to get information without doing the real work of verification. Just understand you are handling a very sharp two-edged sword. Don't get cut, pay attention to the details, and be willing to invest the time and effort to verify.

So, let's look specifically at the ease with which a bad guy can create an internet hoax and how much money can be made from it.

Let's take YouTube as an example. Anyone can create a YouTube channel for free. You simply set up the account, create a name, and you are off. Next, you begin to add video content. The idea is to get as many eyes on your channel as possible. So, the more interesting your name, content, and keywords are, the better. Also, remember the idea is to create an audience and appetite for the videos you post on your channel. So identifying a target audience such as those interested in the UFO topic is ideal.

Next, you can choose to "monetize" your video posts. At no upfront cost to you, YouTube will place advertisement attached to your videos. As people click on your video and on the ads, you will receive compensation. So, obviously, the more people that subscribe to your channel results in more clicks on the videos and clicks on the ads. That's how it is done. The driving factor is clicks on your content. So, truth and integrity have nothing to do with it. In fact, the more attention grabbing, sensationalistic, salacious, outrageous, the videos are the better!

You can see how this has created a dearth of bad videos, bad websites, and bad guys hoaxing for personal fame and money. So, as you research the UAP topic, be well aware of the possible, even likely, false nature of photos and videos on the internet today. Use common sense and verify, verify, verify!

Let me also just say that, as a person interested in the UAP topic, science fiction, movies, books, and various forms of entertainment, that even some of these fake videos can be entertaining. I don't think there is anything wrong with that. Many people are talented, creative, and share a love of the topic and express it through art. I am totally OK with that, EXCEPT if it is passed off as real.

The last thing I'll say on the topic is that there are people who so much want the ET hypothesis to be true that they easily believe things they should not. There are websites that post a combination of both possibly valid and possibly hoaxed videos. I wouldn't necessary call them hoaxers, but I would call them gullible or not entirely diligent in verification efforts.

So again, as an amateur researcher, always beware. In fact, assume what you see and hear is not valid. Then research deeply until you are convinced otherwise by good data.

Note to Self

Hoaxers do a disservice to all by creating a "boy who cries wolf" effect, thus causing all photo and video evidence to be doubted.

Chapter 4

The Scientific Method

One facet of the scientific method could be described as the continual process of formulating a theory based on known facts. Then, as new facts are discovered, the theory is either confirmed or revised based on these new facts. Because of this process, we know much about some UAP, but there is clearly much we do not know. Only by a thorough, sustained, and properly applied scientific study will we learn more.

Make no mistake, even though the media and popular culture often portray UAP as a joke, there are some credible and serious people, performing credible and serious science, based on credible and serious evidence.

Unfortunately, due to the culture of ridicule created around the issue of UAP, it is difficult to get mainstream scientists or institutions to commit to such a study. Scientists succeed or fail based on their personal reputations, credentials, and ability to attain grants for research by the government or universities. The system makes it difficult for them to place themselves at such professional risk. However, there are more and more courageous individuals and institutions willing to take a stand and engage in a scientific study of UAP phenomena. I already mentioned the likes of UFODATA, AATIP, TTSA and a recent group (SCU) the Scientific Coalition for UFO's. It is only a matter of time before a breakthrough is reached.

A classification system for UAP cases was established by Project Blue Book's Dr. J. Allen Hynek to help determine the qualitative nature of case reports. This has a direct bearing on the amount of data collected on a sighting, which leads to a greater ability to accurately identify the UAP. This system of categorization of the qualitative nature of a sighting is known as the "Close Encounter Scale."

Hynek felt that sightings within 500 feet of an object minimized the risk of a misidentification and felt that reports that fit these criteria were especially valuable. Again, his concern was to develop a system of rating UAP reports in a way that could yield data for use in the scientific method.

He classified reports in the following manner:

Close Encounter of the First Kind	Visual sightings within 500 feet of an unidentified flying object.
Close Encounter of the Second Kind	Visual sightings within 500 feet plus the accompanying of physical evidence.
Close Encounter of the Third Kind	Sightings within 500 feet of "occupants" in and around the UFO.

Many people consider a UAP sighting as seeing a momentary light in the sky, or momentary object in the daylight, but fail to realize that there is a difference between a "sighting" and an "encounter." Many cases involve minutes or hours of observation. Many even involve actually interacting with the UAP.

There have been cases of military fighters engaged in aerial maneuvers (Tehran 1976), even dogfights (Peru 1980) including the firing of weapons upon the UAP. In fact, people have been injured or killed (Cash/Landrum Case, Texas 1980), (Capt. Mantell, Kentucky 1948) as a result of an encounter with UAP. We will examine the details of these classic cases in a later chapter. But, suffice it to say, yes, Unidentified Aerial Phenomena is a physical reality.

Generally, the scientific method does not produce huge leaps in our understanding. Such improvements in scientific understanding is typically the result of a gradual process of development over time. Scientific models vary to the extent in which they have been experimentally tested and in their acceptance in the scientific community. In this way, explanations become accepted over time as evidence accumulates on a given topic, and the explanation is better than its alternatives at explaining the evidence. Often the explanations are altered over time, or explanations are combined to produce new explanations or theories.

After nearly 70 years of official and unofficial research of UAP, much has been learned. We have built upon prior knowledge and experience from the scientific method being applied to the study of UAP. For example, we have learned a great deal about what UAP is not. However, *what it is* has eluded us to this point. But there is good reason to continue its study—to test our theories and formulate new ones.

Theories can also be affected by other theories. For example, thousands of years of observations of the planets were finally explained by Newton's laws. However, later these laws were then determined to be special cases of a more general theory of relativity by Einstein, which explained both the exceptions to Newton's laws and predicting and explaining other observations such as the deflection of light by gravity. So, in certain cases, independent, scientific observations can be connected to each other.

The scientific method requires intelligence and imagination, an open mind. It is not a fixed set of standards and procedures to follow, but is rather an ongoing cycle, constantly developing more useful, accurate and comprehensive models and methods.

For example, when Einstein developed the Special and General Theories of Relativity, he did not refute or discount Newton's *Principia*. On the contrary, if the astronomically large, the quantumly small, and the extremely fast are removed from Einstein's theories – all are phenomena Newton could not have observed – Newton's equations are what remain. Einstein's theories are expansions and refinements of Newton's theories and, thus, increase our confidence in Newton's work. So, we must be open minded to all possibilities so that when we are presented with new ideas, we do not dismiss them out of fear, ignorance or pride. In other words, go where the data leads us!

So, here is a typical step-by-step scientific method model that we could apply to our research of the UAP phenomena:

Define a question

What is UAP? This is the burning question we all want answered. Not what could it be, not what it isn't, but *what is it?* We all want identification of UAP in a scientific, factual, and defendable way. We want definitive proof. So, we need to make sure we are asking the right question, the right way, to yield that result.

For example: "Do you believe in UFOs?" You would think the United States Air Force term "Unidentified Flying Object" was definitive enough, but it has led to a great degree of confusion. This question has come to mean "Do you believe in aliens?," which is, in fact, not the right question!

So, in your research of UAP phenomena, make sure you ask the correct question so you can determine the right answer.

Gather information

Observe and record evidence. To answer our question in a scientific and definitive way, we need evidence. Since these pesky UAP seemingly have a mind of their own, we cannot, at our convenience, simply bring them into a lab and observe them. It's a lot harder than that because it appears UAP aren't ready to give up their identity that easily. So, what do we do? Observe them when we can. This is usually through the incidental capture of them on photographic, video, or radar imagery for analysis. We also have trace evidence that can be analyzed and scientifically identified. There is also witness testimony that can be recorded and analyzed as well. In fact, the next entire chapter will be dedicated to the forms of evidence that can yield scientifically provable results.

Form a hypothesis

UAP are real and a physical phenomenon. So, as we have made observations, analyzed physical evidence, and reviewed witness testimony, we no doubt have begun to form some opinion about what the evidence and analysis is indicating. This is a hypothesis. In other words, this is what we think may be happening. It is not as complex as a theory, but the beginning of formulating one based on what the evidence is suggesting.

For example, we may have analyzed a video of a craft that clearly does not visually match any known craft and its flight profile does not match that of any known propulsive technology. But it appears to be a "nuts and bolts" physical object, demonstrating real physical activity. We may then conclude or hypothesize, if it is not a hoax or misinterpretation of something conventional, that this is a real and physical phenomenon. That is a reasonable, scientific conclusion to make as part of the scientific method.

Test the hypothesis

Perform experiments and collecting data in a verifiable manner. As mentioned above, these things don't exactly cooperate by making themselves available for inspection. So, how can we test our hypothesis? Perhaps we can compare the propulsive performance of known technology to that of the UAP as captured in the video. So, we know that current propulsive technologies in use today boil down to thrust and lift accomplished by a variety of technologies such as internal combustion engines driving propellers or rotors, jet engines, rocket engines, or any combinations of these. If the observed craft did not display flight characteristics of these known propulsion systems but rather demonstrated silent, anti-gravitational maneuvers, then we have at least tested our hypothesis with available evidence.

Analyze the data

Once measurable data is available, mathematical, chemical, relational, statistical methods of analysis can yield a great deal of information. So, suppose there is an alleged landing site of a UAP.

What information may be available for analysis and what can it tell us about the hypothesis we have formulated?

Perhaps tree branches are broken above the landing site. Wouldn't that suggest a physical object crashed through them on the way down to the ground? Could there be radioactive or chemical traces left on those branches from the body of the UAP or from its exhaust? Would it not be practical to perform a chemical analysis to confirm this? How about indentations in the soil from the landing gear of the UAP? Would not a mathematical analysis of the dimensions of the impressions in the soils say something about the weight of the craft? On a statistical basis, a comparison of this data to other cases may yield a pattern that adds to our knowledge and may reinforce our hypothesis. So, measurable data and its analysis can confirm or cause us to reevaluate our hypothesis.

Interpret the data

So, we may be well on our way to forming a theory once we draw conclusions to validate our hypothesis based on our interpretations of the data. We may be able to say we can confirm UAP is a real and physical, phenomena because this evidence adds up to a conclusive identification of the UAP. Accounting for all of the known factors, in the correct context, the only conclusion that makes sense is that this is an Unidentified Aerial Phenomena. It may not reach the heights of absolute proof, but it is a strong evidential theory based on a process of elimination.

Publish results

Write a report, white paper, article, or book after you are confident with your conclusions and have the defendable data to back it up. This allows others to review the work you have done. This would be a scary proposition because of the risk of others finding a fault with your theory, but in the end, we all want the truth. It is worth the risk and the scientific method demands it.

Retest

If the theory stands up to the peer review, then others will need to test your theory using the scientific method which is simply to repeat the iterative cycle in this step-by-step method from testing the hypothesis, analyzing the data, interpreting the data, and doing so again until a matter is proven.

This is the way in which we can yield quantitative and qualitative data that can be taken seriously and will lead to a defendable theory or identification of UAP.

I understand you personally are unlikely to be involved with the investigation of a UAP incident. However, you will be reading books, reports, documentation, and other resources as you engage in your personal study of the subject. So, be aware and look for those researchers, organizations, and others that have performed investigations in a professional, scientific manner. These will be among the most reliable and defendable cases.

Note to Self:
There are some credible and serious people, performing credible and serious science, based on credible and serious evidence.

Chapter 5

Forms of Evidence

While most of UAP sightings can be attributed to misidentification of known phenomena, there are many cases that cannot. As mentioned in an earlier chapter, on average 21.5% of cases are attributed to unknowns. Remember, not because there is a lack ample evidence to make an identification, but because there is ample evidence to eliminate any known phenomena!

So, what are the forms of evidence that provide data and information that will help yield an identification of UAP or eliminate prosaic explanations? Let's consider several.

Eye Witness Accounts

People from all walks of life report UAP. These reports are among the most useful forms of evidence and data because they are the basis for understanding the circumstances and context of a sighting or encounter. This is where an investigation begins, the report of what was seen or experienced. So, what must we consider when it comes to the value of eye witness reports?

An eyewitness report can be enough to make an identification of UAP. Often, sightings are misidentification or misinterpretation of prosaic things such as balloons, lanterns, meteorological formations, aircraft, and the list goes on.

One reason these things can be mistaken is the result of an interpretation of an eye witness who is not expecting to see such a thing in that time and place. It is out of context, or it is ambiguous enough that when combined with the timing that an explanation is hard to imagine.

Of course, there are other people who have poor eyesight, an active imagination, are under the influence of drugs or alcohol, or in some other way lack credibility. However, with an experienced, investigative effort, the identification often is possible.

I love the quote from UFO research pioneer Stanton Friedman who quipped in his book *Flying Saucers and Science*, "It is not always the town drunk making a UFO report at 4:00 in the morning." This leads to an important thing to consider when determining the credibility of a UAP eye witness. Is this person credible in other areas of life? If so, then it is more likely they have not misinterpreted the sighting as described above. That's no guarantee, but it helps one have confidence in the report.

So, what would you consider a reliable or credible eyewitness? One might be someone you know personally. Someone with whom you have had dealings who have exercised good judgement. Another may be someone who is in a respected position in society. It may be someone who has authority based on technical skill, expertise, and judgement.

Let's consider a few examples of who are considered reliable UAP eyewitnesses by profession.

Police Officers are persons who have been highly trained, certified, engaged in continuous training, are constantly monitored for physical and mental health in the performance of their duties. They are placed in life and death situations on a daily basis. They are constantly required to make observations and the correct decision under pressure, to exercise good judgement about their surroundings, including the behavior of people around them. Though not trained as scientists or engineers, they must be among one of the most credible observers and witnesses around.

Pilots, whether they be military, commercial, or private are also particularly credible eye witnesses of UAP. In fact, since aircraft, flight dynamics, flight operations, take-off and landing, navigation, and meteorological conditions are their business, their life, they may be the best eye witnesses of UAP! Can you imagine as a pilot having the lives of hundreds of people in your care, dependent on your skills and knowledge? Can you imagine what they have seen in the skies over the thousands of hours and millions of miles of flying? It's been estimated that not even 10% of UAP sightings have been reported due to the ridicule and persecution that would come upon pilots if they did indeed report UAP.

Scientists, engineers, and technicians are also typically good eyewitnesses. Since they have been trained in observation techniques, quantitative and qualitative analysis, data collection and review, and all that comes with the scientific method discussed earlier in this book, they must be considered credible witnesses. Although, they are dis-incentivized to do so by the same ridicule and persecution that pilots experience as well.

Government Officials also make credible witnesses, but not due to any technical acumen, rather because they are at risk of ridicule, persecution, even loss of career. So, if they are brave enough to make a report at such risk, you can be sure they are sincere in their report. They may be mistaken, but certainly not making a glib report. They likely weigh the heavy consequences that could come their way before making any report.

Military Personnel are also credible eyewitnesses because of many of the reasons, police officers, pilots, scientists, engineers, and technicians because each of those professions are within the military. Even non-technical military members are well trained in observation. They too must exercise good judgement and soundness of mind. Moreover, they are quite likely to observe UAP in the execution of their duties within the military. UAP seem to be quite interested in all things military, including nuclear activities. So, the incidence in seeing them is very high.

So, as you research UAP, consider this form of evidence, namely eye witness testimony, with caution. Consider the source and its credibility. That is not to say an average Joe, or even the town drunk is incapable of seeing and reporting genuine UAP credibly. Nor is it a guarantee that one of the above are always sound in mind and good judgement. Like any good investigator, use multiple sources, consider all the evidence, and again, make sure your BS meter is well calibrated!

Trace Evidence

Trace evidence is an extremely valuable form of evidence because it allows for the collection, analysis, and data driven conclusions about a UAP event. In other words, it is physical evidence or proof that, at a minimum, something physical occurred. This is something that can be utilized in the scientific process. It can be measured, aggregated, and otherwise used to develop a hypothesis.

Trace evidence can come in many forms. For example, it may be chemical, radiological, or electrical effects, it may be damage to plants, animals, even human beings.

An excellent example of the value and application of UFO trace evidence is embodied in the work of Ted Phillips—a protégée of Dr. J. Allen Hynek. Ted specialized in trace evidence. He compiled a catalog of trace evidence types and cases. I have recreated below an actual summary of a specific case from the 1970s which reflects the trace evidence collected, the process of analysis, and the conclusion drawn from the analysis.

This recreated data is from a PDF of the typed documentation found on pages 137 and 138 of the report titled, "Physical Traces Associated with UFO Sightings," compiled by Ted Phillips and published by the Center for UFO Studies in 1975. The document is posted on www.cufos.org. The analysis is of the trace evidence in the Delphos, Kansas landing case on November 2, 1971.

Appendix

SUMMARY OF THE SEVEN SOIL ANALYSIS REPORTS

1. Standard Soil Analysis as used for Agricultural Purposes. Agricultural Experiment Station, Utah State University.

These two soil samples (from the ring and a nearby control sample) are markedly different. That from the ring is extremely difficult to wet with water, and is much lower in Ph (that is more acidic) and higher in soluble salts. It is also significantly higher in Mg (magnesium) and K (potassium).

2. Energy Dispersive Analyzer (EDAX) with Scanning Electron Microscope. Materials Science Department, Northwestern University.

Sample I (hard-packed ring soil) contains about five to ten times more calcium than Sample II (soil outside of ring.)

3. Soil Fertility and Suitability Tests Performed on Samples Supplied by Ted Phillips to Stanton E. Friedman to Agri-Science Laboratories, Hawthorn, CA.

Characteristic	Ring	Normal
PH	6.0	6.4
Salinity (mMHOS/cm2)	15.0	4.0
Saturation%	24.62%	17.5%

Major Elements   (concentrations in ppm available)

Calcium	2400	940
Magnesium	730	87
Potas	1680	940

Trace Elements

Iron	28.0	6.8
Manganese	56.0	5.2
Copper	2.48	1.0
Zinc	20.0	0.18

The white material in the soil "was defined by Agri-Science as a non-organic water-soluble material." (Stanton T. Friedman)

4. Seed Germination Tests. (Unspecified University)

"The ring soil produced mush less seed growth than the control soil." (Stanton T. Friedman)

5. Spectrochemical Analysis and Quantitative Estimates. Commercial Testing Laboratory.

Soil from the ring sample found to be markedly higher (0.5 to 5% compared to 0.2% to 2%) than nearby soil in concentration of calcium, and less significantly higher in concentrations of trace elements, magnesium, manganese, lead, boron, silver, tin and chromium. Ash content of the ring soil sample was significantly lower (83.7%) than in the nearby soil sample (91.07%).

6. Infrared and Raman Spectroscopy, Electron Spectroscopy, and Chemical Solvency Tests. Chemistry Department, Northwestern University.

> The results of these tests are consistent with the hypothesis that the ring soil is coated with a hydrocarbon . . . a material that was later removed from its surface at 100 degrees C under the conditions in the ESCA spectrometer. The hydrophobic nature of the ring soil . . . was affected when ETOH (ethyl alcohol) then water was added. Apparently, . . . something is washed away so that surface properties of the soil are changed.
>
>   7. Scanning Electron Microscopy and Other Methods by un-named Laboratory.
>
> Soil particles of about ¼ micron in diameter (about 1/4000 of an inch thick) were found to be coated with a layer 1/50 micron thick of a low atomic-weight material in which some small particles were embedded. Unique icicle-shaped crystals about 0.1 to 0.05 microns were found, each with a dozen or so spherical globules of higher atomic weight adhering to the surface of the crystals. A crystalline structure of a previously uncatalogued diffraction spacing was found in the particle-coating material. These abnormalities were found only in the soil near the surface of the ring.
>
> One of the physicists commented "Incidentally . . . when I unwrapped the soil samples . . . we were both hit with a brief coughing spell. Apparently, the dust has some irritating quality, and we're using respirators in handling the stuff from now on." Publication by the laboratory of the complete analysis is pending. (CUFOS)

So, this case evidence was largely chemical in nature. It also considered the effects on the plants and soil at the point of contact with the UAP. There were additional forms of evidence including the eyewitness testimony, the effects on the people who came into contact with the soil, and the long-term hydrophobic effect on the ground. I wanted to include this so that, in your UAP research, you will be able to recognize genuine, scientific, and high-quality research and trace evidence.

Photographic, Video, and Radar Imagery

I would love to be able to say that a photo, video, or radar image could provide not just evidence, but proof of UAP! But, alas, it is just not so. The sad truth is that these forms of media are easily falsified. In fact, just let me repeat what I mentioned in an earlier chapter:

Now, more recently, with the invention and availability of computer software, an individual can create some convincing Computer-Generated Imagery (CGI) that, to the untrained eye, may appear valid. Although, with scientific scrutiny, these too can be identified as fraudulent. Sadly, today many people will see a photo or video on platforms such as YouTube, and not take the time to research and determine the validity of the photo or video. Even when you do, it is also very difficult track down the source to verify the back-story. So, the internet is chock full of photos, videos, and websites that are complete fabrications. So, researcher beware!

Now, we cannot summarily dismiss all photo, video, and radar imagery. In fact, if we ever do identify UAP, or if it is disclosed by a government, this imagery will no doubt play a part in the authentication of the UAP. It will just depend upon the circumstances and who is presenting it as evidence.

One example of a *photograph* that is generally accepted as an actual photo of UAP is the September 4, 1971 Lago de Cote, Costa Rica case. The photo was taken by a government mapping plane. The circumstances were ideal for knowing the provenance of this photo. This was taken in the execution of the mapping team's official mission. There was not only the pilot, but also the aerial photographer, a geographer, and a topographer. They were using high quality photographic equipment, they knew the date, time, altitude, the scale of the imagery, all the factors necessary to confirm the actual presence of the object. This also gave them the ability to determine its size. It is a sharp image against a dark background, which provides clarity not usually associated with many UAP photos.

With such good data, both Jacques Valle and Richard Haines—highly credible researchers—have been able to complete an exhaustive and in-depth analysis of the photo concluding it was an actual image, not hoaxed, with no apparent prosaic explanation—a genuine UAP!

One of the most compelling video images released to the public is the "Tic-Tac" gun camera video shot from an F-18 Super Hornet in 2004. This video was taken during the USS Nimitz aircraft carrier fleet training exercises prior to deployment in the Persian Gulf during the Iraq war. The investigation of this case is occurring at the time of writing this book. However, many details, testimonies and physical evidence is in the process of being investigated. The lead investigation for the Department of Defense was a program entitled :Advanced Aerial Threat Identification Program" led by Lou Elizondo, who retired in 2017 to become a lead investigator for the "To the Stars Academy of Arts and Science (TTSA), since the Department of Defense closed the AATIP program. The investigation is continuing under the auspices of TTSA. What makes this video so significant is that it contained a great deal of verifiable data such as thermal, optical, and radar imagery generated by a multi-million dollar aircraft, highly skilled pilot and radar specialists. Its provenance is irrefutable and has been confirmed by the Navy as we speak. There are two additional gun camera videos called the "Go Fast" and the "Gimbal" videos that were taken during carrier group exercises on the east coast in 2014 and 2015, under nearly the same circumstances.

I personally consider these videos as some of the most intriguing evidence of UAP. You can find these videos on YouTube, or many UFO research sites.

Finally, a word about *radar imagery*. I consider it a very good form of evidence largely because the imagery comes from a scientific instrument, in a closed and controlled environment with a high degree of technical skill to operate. In other words, it is not prone to hoax. There can be misinterpretations of the data, but the conditions for such are well known such as temperature inversions. However, skilled and experienced operators are quite adept at distinguishing between false images, sweeps, pings, and the like. These scientific or military systems are maintained, patched and otherwise kept up to date. A strong case involving the confirmation of a UAP by radar imagery is the November 17, 1986 Japan Airlines flight 1628 encounter over Anchorage, Alaska. In an amazing encounter of a Boeing 747 cargo plane and a UAP the size of an aircraft carrier for over 50 minutes, the whole affair played out on radar. This is a terrific example of the value of radar imagery as evidence for a UAP case. Let's walk through this.

The JAL flight was transporting a load of wine from Paris, France to Tokyo, Japan. During what was a routine flight for the Boeing 747 jumbo jet configured for cargo transport, a UAP approached the flight over a remote part of eastern Alaska, USA that evening at approximately 5:00 p.m.

Two large UAP approached on the 747's left side a few thousand feet below, but soon rose to within 500 feet of it. This caught the attention of the pilots and caused great concern. At one point, the two craft pulled directly in front of the plane and applied some form of thrust that lit up the 747's cockpit with overwhelming bright light, and it actually heated the cabin to the extent the pilots felt heat on their faces!

Minutes later, a third UAP approached from behind and then moved next to the plane. The pilot estimated its size to be equivalent to an aircraft carrier!

The JAL aircraft was in contact with one military and one civilian air traffic controller during the event. At one point, the ATC said the 747 had a "bogey in trail." The pilot requested a course change, which resulted in a full 360 degree turn in an effort to "shake off" the trailing UAP. After the flight got closer to the city lights of Anchorage, the UAP pulled away and disappeared in the night skies.

As remarkable as a close encounter as that was, the ensuing investigation and surrounding events are fascinating in themselves. A report was made by the captain, and it eventually landed on the desk of John Callahan, the FAA Division Chief of the Accidents and Investigations branch.

Once all the materials were forwarded from Alaska to New Jersey for the review, Callahan and others ingeniously overlaid the voice recordings of the pilot and air traffic control conversations with the radar imagery matched in real time. They then videotaped this compellation of voice and radar data to form a very compelling record of the close encounter.

The investigation proved to be fascinating itself. The compelling presentation was made to members of President Ronald Reagan's Science Committee, FAA officials, and officers from the CIA. After a thorough discussion of the voice and radar data, technical specialists determined that these events and testimony of the pilots, air traffic controllers, and even another nearby flight confirmed the presence of UAP as described. In fact, when Callahan asked the CIA officer what he thought this was, the officer stated, "It's a UFO!" He then dryly instructed everyone to keep this quiet, that "this never happened," and to send all the records to the CIA.

Callahan complied, but had a second copy in his office that they never picked up. So, when he retired, he took it with him. The reports, voice and radar recordings, everything! Thankfully so, or we would not have one of the best radar-evidenced close encounter cases of all time out in the public domain!

As you can see, radar data is difficult to refute. When combined with the other forms of evidence, such as eyewitness testimony, trace evidence, video and photographic imagery, and even audio files, a very solid investigation and credible and high-quality conclusions can be made. There are many such cases just waiting for you to research. Take advantage of these and learn what a good case looks like, so you can develop into a good researcher.

Note to Self:
The higher the quality of evidence, the higher the quality of conclusions!

Chapter 6

Classic Cases

There are many well documented cases in history that provide sufficient data to determine the fact that there is a very real and genuine phenomenon occurring. I am talking about nuts-and-bolts, physical, metallic objects as real as the airplanes flying over our heads! I'm talking factual, provable government interest and participation in UAP phenomena. I'm talking physical evidence in the form of trace evidence, photographic, video, and radar imagery. I'm talking very credible eyewitness and whistleblower testimony. Admissions to its reality by government officials and the documentation to back up that assertion. Yes, all of this exists and is available to you, even as an amateur researcher. Thanks to living in a free country, the availability of literature and the access provided by the internet, any of us can dig in and find true, factual, and satisfying answers!

So, in this chapter, I am going to share with you a variety of cases that I recommend you should research as you begin your journey. They contain many of the attributes of a classic case. Each case has its strengths and weakness. But, when you consider them as a body of evidence, they prove to me beyond a shadow of doubt the reality of this phenomenon. I hope you find them meaningful as well. I am only going to highlight them in a way to lead you in the right direction, but I will leave you plenty to discover on your own. I highly recommend these be among your first cases to delve into.

I've included several cases, each for the type of evidence that makes them a strong case. For example, some cases have strong photographic evidence, some have compelling trace evidence, some are interesting because of who the eyewitnesses were and the testimony they gave. The point is, each case has strengths and weaknesses when it comes to evidence.

Each case also stands on its own merits. So, while one case has different evidence, each can be convincing for various reasons. I hope this will become clear as you review these classic cases. I am going to relate them to you in chronological order, not in importance or quality, because, in their own way, each is important and credible.

Just a word about the phenomenon from a historical perspective first. In my research, I noticed that the 1940s through the 1950s were a very rich time period, not only for the appearance of UFOs, but the response by the military, the government, and the public. This is considered the point when the modern UFO enigma began.

It took time for people and organizations to get their heads around what was going on and what it meant. Mind you, this was during World War II and during the cold war. If you were not alive at that time, you may have a hard time grasping how significant a mindset that made for the nation. Virtually everything going on in society was looked at through the lens of war, peace, rebirth, technology, and fear. There was very real concern that this UFO enigma was a Russian construct to generate fear and disorientation to disrupt America.

What this mindset did for ufology was create a public, accessible body of information available for us today. The newspapers were full of reports and accounts of UFO sightings: Official government and military studies, reports, policies, documents, books, news conferences, and even open and honest engagement with interested individuals and organizations occurred. Although, that began to change, and secrecy became the official policy, which continues to this day.

Even though the societal, governmental, and military view has changed over the years, one thing has remained consistent. Those pesky UAP will not go away!

So, let's begin with a very interesting mass sighting that occurred in February of 1942. It is known as the Battle of Los

Angeles.

Battle of Los Angeles, 1942

Only a few months after the United States of America entered World War II, an incident off the California coast occurred involving the US military and the community at large with a perceived attack upon the coast. This comes on the heels of the December 7, 1941 attack on Pearl Harbor. The military has since described it as a case of war nerves. A weather balloon reportedly got loose and drifted down the coast. Jittery artillery battalions allegedly were confused and unleashed a barrage of some 1,400 rounds of artillery over a several hours to bring down this "balloon." Of course, no downed balloon was recovered.

There were reports of a saucer-like craft all down the Pacific Coast to Long Beach. Individuals died from heart attack and car accidents. Shrapnel from those artillery rounds littered the beaches and neighborhoods in the path of the shelling. A car was damaged by a round of artillery, and other minor property damage was reported. Eyewitnesses were confused why a "balloon" would not be popped by this barrage of artillery. The official explanation did not hold water with many people.

Of course, there is a very well-known photograph taken by the *Los Angeles Times* newspaper. You will find resources to research this case on the internet and in books. I highly recommend David Marler's presentation on the Battle of Los Angeles, which can be found on YouTube. He did an excellent job pulling together the facts, and has presented them in a very compelling way.

Roswell, NM, 1947

Roswell is the case of all cases. This event, the subsequent reaction of the military, and the discovery of all the eyewitnesses and the recording of their testimony by the likes of Stanton Friedman, Tom Carey, and Don Schmitt has made this "THE" UFO case of all time. Other publications such as *The Day After*

Roswell by Corso and Birnes adds controversial thoughts to this case.

I'm sure you are already aware of this case, but let me share a brief summary, as described on the MUFON website:

> "Roswell UFO incident took place in the U.S. in June or July 1947, when an airborne object crashed on a ranch near Roswell, New Mexico. Explanations of what took place are based on both official and unofficial communications. Although the crash is attributed to a secret U.S. military Air Force surveillance balloon by the U.S. government, the most famous explanation of what occurred is that the object was a spacecraft containing extraterrestrial life. Since the late 1970s, the Roswell incident has been the subject of much controversy, and conspiracy theories have arisen about the event.
>
> The United States Armed Forces maintains that what was recovered near Roswell was debris from the crash of an experimental high-altitude surveillance balloon belonging to what was then a classified (top secret) program named Mogul. In contrast, many UFO proponents maintain that an alien craft was found, its occupants were captured, and that the military engaged in a massive cover-up. The Roswell incident has turned into a widely known pop culture phenomenon, making the name 'Roswell' synonymous with UFOs. Roswell has become the most publicized of all alleged UFO incidents."

There have been many books, television programs, articles, and even movies about the Roswell event. There is more information than you can consume on this the world over.

However, I recommend you start with the books, *Witness to Roswell*, by Tom Carey and Don Schmitt, and, *Crash at Corona*, by Stanton Friedman. There are many more, not all of which I have read. But this is a good start.

Just remember, this case has been the subject of much debate. Debunkers, including the United States Air Force, have tried to discredit this at every turn. There is much lore that has surrounded this case as well. So, buyer beware! There is a lot more to this case than most people are aware. So, this is worth a deep dive, and should be among one of your first case studies.

Washington, DC, 1952

This case was one for the history books. Imagine unauthorized aircraft flying over the nation's capitol, toying with scrambled military interceptor jets. This got the attention of the military, the President, the media, and the public in a big way. Headlines in newspapers across the entire nation alarmed the public in a way not felt since WWII.

It all began on a summer's day in July of 1952. In fact, for two successive weekends, the "Washington Flap" UFO reports streamed in of unidentified objects flying over Washington National Airport (now known as Ronald Reagan Airport).

Airport radar operators saw blips on the radar that they knew to be anomalous due to their flight behavior. The control tower staff also confirmed the radar blips, but also confirmed sighting bright lights that zipped away at incredible speeds.

Eventually, fighter jets were scrambled from nearby Andrews Air Force Base. The objects would toy with the jets, disappear, and reappear over the course of several hours.

Various radar systems from several towers, both military and commercial, recorded the UFOs' movements over many hours and days. This story was reported by local, national, and international news outlets. The military was in an uproar over this incident. The Pentagon, FBI, CIA, and the White House all were concerned and made some attempts to figure it out. In fact, this event led to the largest military press conference since WWII.

I recommend this to be one of the first classic cases you study. It is one of the foundational cases that engrained the current government and military public repudiation campaigns regarding UFOs and UAP.

Make sure you watch the movie *UFO*—the Al Chop story. This is one of the best docudramas made about a specific UFO case. It was made with the very traditional 1950s authoritative mindset. It touched on the UFO themes of the day, which was the heyday of UFO sightings and reports. It also contained actual footage of a few cases, which was being made public for the first time. There were Project Blue Book cases that had excellent witness testimony and physical evidence. In addition, Richard Hall refers to this event in his 1954 book, *The UFO Evidence*.

Soccorro, NM, 1964

By 1964, mostly just sightings of objects in the sky had been reported. However, in that year, an encounter occurred that made the Air Force consider the possibility of beings piloting these so-called UFOs. It also caused an entry into the official records (Project Blue Book) of an element not yet recorded; namely aliens!

It was on the afternoon of April 24, 1964. New Mexico state highway patrol officer Lonnie Zamora was on duty and had just begun chasing a speeding Chevrolet through town and onto State Highway 85. They began to go through a hilly area when Officer Zamora heard a loud roar and caught sight of a strange flame-like light in the sky in the near distance, no more than a half mile away. This was strange enough that he decided to abandon his chase of the speeder and to investigate.

As he got closer, he noted the flame was a bluish-orange and appeared stationary. He pulled off the highway onto a dirt road to get closer to the source of this light. He had a little difficulty getting up some of the hilly dirt roads in his patrol car, but finally came up to the top of a hill on his third try, discovering a shiny object down below him, maybe 150-200 yards, that he first interpreted as a possibly overturned car.

According to Robert Emenegger in his book *UFO's: Past, Present and Future*, Zamora immediately called it in to the dispatch at the Sheriff's Office. The officer on duty took the call and remembered, *"The time was about five forty-five I recall."* He said, *"Socorro-Two, possible ten forty-four, an accident. I'll be ten-six."* This meant the officer was leaving the vehicle and radio to go investigate the overturned car. The officer said, *"I could hear that he had stopped the car and got out."*

Zamora described what he did next: "As I approached the scene, I thought it was an overturned car-on end sort of. As I got a look at it, it was a shiny-like object. The object was like aluminum—it was whiteish against a mossy background, but no chrome. Seemed like a football shape. I saw two pair of overalls."

Of course, what he meant by that was he saw two "people" walking outside of the vehicle. He began walking down the hill to go help, but he recalled one of them looking directly at him, and both reacted as though they were startled by his presence.

Suddenly, they ran back to the "car." and that loud roar began again, not jet-like, but beginning with a very loud, low-pitched humming noise escalating into a very high-pitched hum. Then came the flames. Fearing for his safety, he turned around and ran for the car, even bumping into it, falling down, and knocking off his glasses. Zamora recalled, "I ducked down and ran over the hill." He related the fear he felt in this way: "I was scared of the roar. I turned back, at the same time covering my face with my arms. There was no roar now. I looked up and saw the object moving away from me. When the roar stopped, I heard a whine and then complete silence."

Zamora called it in to the police station and, as he spoke to the operator, he watched the object lift up slowly, then disappear rapidly into the distance. He sounded hysterical to the officer on the other end of the line.

Sergeant Chavez responded to the call and, upon arrival, noted, "When I arrived, Zamora was sweating and white, very pale.

I went down to where the object had been. I noticed the brush was burning in several areas. I could see tracks on the ground. The object had left four, perpendicular impressions in the ground. I noticed smoldering bushes, but they felt cold to the touch. I knew Lonnie had seen something, the proof was right there. Before I arrived, he had made a sketch of an insignia he saw on the side of the object. I secured the area and notified the local military authorities."

Quintanilla, the Project Blue Book Air Force officer in charge related, "We were later informed that an FBI agent had been on the scene before us. By the time we arrived on the scene, the area was crowded with spectators and newsman. They were all over the place. The area had been roped off but still the marks on the ground had been disturbed. I directed Dr. Hynek to Socorro to get additional information."

Dr. Hynek recalled, "When I arrived there off the plane, flashbulbs began popping and I was asked for one interview after another, but I had just arrived and didn't know anything! I finally got the newsmen out of my hair when I finally interviewed Officer Zamora, who was not very talkative. I questioned an Art_____, of the FBI and several other who had been on the site within the first few hours after the sighting about the freshness of the tracks. They were all of the opinion that the tracks were indeed fresh. The incident was very interesting. It seems to differ from practically all the earlier cases we investigated from one standpoint. The vehicle left pod marks. There was an insignia observed by Lonnie Zamora on the side of the craft, which I will go into in a minute, but the insignia was very interesting and unidentifiable, not American nor Russian and last of all, the observation of these two people in some sort of suit."

The insignia was an arrow pointing up, an arc line above it, and a simple horizontal line below it. I will let you discover the drawn image by Zamora on your own as part of your personal research. It is fascinating to contemplate, isn't it? This is the first, officially reported insignia on a UFO! In addition, this is the first case with "people" associated with an object. Zamora described them as "kids or small adults" in white overalls. That is fascinating to contemplate as well.

So, what was the Air Force's estimation of the situation? Quintanilla shook his head, recalling the incident like it was "the one that got away." "Well, we went over the case pretty completely. We had soil samples tested at the Air Material Lab. The burned bushes were analyzed for propellant residue. We measured the distances between pod marks. There were even footprints left on the ground. My first reaction was that it was a lunar test module from NASA or the Air Force. That seemed the only logical explanation."

He added in a slow, solemn delivery, "This was probably the best documented case in the Air Force files and I checked it out everywhere. All the way up to the top, even the White House Commend Post, and nothing! I wish I could put my finger on it, although I'm not convinced that it was anything extraterrestrial."

The final disposition card in the Blue Book file reads, "Initially believed to be observation of lunar module type configuration. Efforts to date cannot place vehicle at site. Case carried as UNIDENTIFIED pending additional data."

Interestingly, about four months after the incident, Dr. Hynek returned to Socorro to follow up and write in his report. "The object of this visit was to obtain an overview of the feelings and opinions in Socorro about the Zamora sighting of April 24th, after several months have passed, and to find out if the principals had any after thoughts or changes they wish to make in the story, how they were now regarded by the town folk, and what if any, was the official opinion."

Emenegger further sums up Hynek's disposition of the case in this way: "The net result of the visit, which involved talking one again with Zamora, Sergeant Chavez, Captain Holder, the editor of the local newspaper, and seven other towns people, was much the same as before. Zamora, if anything, is more reticent and withdrawn. The more articulate Sgt. Chaves still firmly believes Zamora's story, and I found no contradictions between his partial retelling of the story and the original telling of his story in late

April."

Hynek continues "Although, I made a distinct attempt to find a chink in Zamora's armor, I simply could not find anyone, with the possible exception of Mr. Phillips, who has a house fairly near the sight of the original sighting, who did anything but uphold Zamora's character and reliability, and I talked with people who had known him since childhood. I revisited the site: the markings are still there, but very much obliterated and this time, I was able to take stereo photographs of the general terrain. I was impressed more than before with the illogical nature of the landing site."

Hynek was so intrigued with this case that he took great care to uncover any hidden or missed clues. He re-interviewed people and returned to the site to take more samples. He concluded with this comment:

> "I visited the site and took more samples, particularly of the sap, and I took a few more pictures, particularly of the dynamite shack (near the site) to show Menzel.
>
> In view of the fact that the prevailing opinion in the town is that what Zamora saw was not the result of hallucination or hoax, but a secret test vehicle, what has become of my suggestion to have this left as an exercise for the students? It would be a marvelous exercise for neophyte intelligence officers."

I must interject my own thought at this point. Hynek himself noted that he was interested and perplexed by this case. He went back and re-investigated. He clearly was not part of a government attempt to cover this up, but he also noted he was aware of the criticism of the Air Force to do just that.

Notice he continues:

> "There is also the opinion expressed in Socorro,

and expressed to me a number of times in the past, by several people (and also by La Paz) that I am merely part of a super smokescreen, as so is (FTD) Air Force and Wright Field, and that the whole Project Blue Book is a grand cover up for something the government doesn't want to discuss. Best way to give the lie to this, of course, is to point out if this were the case, the US government should also have been responsible for sightings in France, Brazil, Spain, and in England. Maybe the US government has gone global!

On that happy thought, I conclude my report.

Sincerely yours,

J. Allen Hynek

P.S. I now have a slightly infected finger from the thorns on the bush that was originally charred. The bush drew blood when I attempted to get some soil samples. Undoubtedly, the finger will now wither away from radiation burns. Unfortunately, I do not have interplanetary Blue Cross coverage!"

It is easy to understand that, at this point, the government was feeling the pressure to explain this and other perplexing cases such as the 1952 Washington, DC flap in July of that year.

In fact, notice the comments from this U.S. Air Force INTERNAL memo. In *UFO's: Past, Present, and Future*, Emenegger recounts:

"Numerous inquiries concerning Air Force investigation and evaluation of the recent UFO sighting at Socorro, New Mexico, 24 April, continue to be received from news media, from the office of the President, and members of Congress. This HQ is unable to respond properly pending receipt of your conclusions. Request appropriate action be taken to ensure that reports of personnel assigned to investigate and evaluate the sighting be completed and forwarded immediately."

So, under this pressure, the Air Force authorized this PUBLIC response by Captain Hector Quintanilla:

"The investigators at Wright-Patterson have not been able to identify or determine what type of vehicle or object Mr. Lonnie Zamora observed on 24 April at Socorro, New Mexico. The object or vehicle demonstrated flight characteristics well within the State-of-the-Art and the sighting cannot be attributed to atmospheric or astronomical phenomena. In this respect, I can categorically state that the object or vehicle observed by Mr. Lonnie Zamora was not an interplanetary space vehicle visiting planet Earth. This case is still open and the investigation is still in progress."

Hector Quintanilla was determined throughout his entire career at Project Blue Book to hang some mundane explanation on each-and-every case, whether true or not. He simply could not accept any explanation outside of his military and Catholic mindset. He personally struggled to accept the possibility of extraterrestrial life in this universe, and it was reflected in the disposition of cases in the project. What a shame!

This case also affected Lonnie Zamora. Hynek stated,

"I came back to the trenchant fact that Zamora was a thoroughly scared person. Chavez has remarked this to me a number of times, that never in his long association with Zamora has he seen

Classic Cases 69

him in anything at all approaching the state he was in when Chavez joined him. Zamora is used to accidents, bloodshed, fights, and murders. We all seem to agree Zamora saw something that really frightened him.

He was simply a cop on duty who relinquished one discharge of that duty (chasing a speeding car) for another he thought more important (investigating the possible explosion of a dynamite shack). His fright was genuine, and his feeling that he had seen something truly unusual is attested by the fact that he asked whether he should speak to the priest first before saying anything about it.

He is a cop who looks as though he could be pretty gruff with some of his customers, and in fact his complaint about the UFO sighting was that it did not allow him to give out his full quota of tickets for the day."

It is interesting to note that at least one other person did in fact see the UFO before it landed in the wash. A tourist traveling on Highway 85 stopped at a gas station and talked to the station manager, Opel Grinder. According to Grinder, the man said he saw an aircraft very low in the sky. Grinder informed him that they do have helicopters in the area, to which he replied, "It was a funny looking helicopter, if that's what it was." He also stated that "it flew over his car, over the gully, and that he saw a police car heading up the hill."

The actual witness was never identified or interviewed. That would have really helped back up Zamora's story. Zamora took more than his fair share of ridicule for the whole thing, and quit the force just five years after the incident. He managed a gas station thereafter, and was very reluctant to discuss the sighting. He even remarked if he were to do it all over again, he would never have reported it. Again, what a shame!

There is more to this story, and I highly recommend researching this truly unique case.

Travis Walton Abduction 1975

The final case I would like to recommend you research is that of Travis Walton. I have not pursued the research of many abduction cases yet. However, even though this is an abduction case, it is unique and captured my attention. Like the Phoenix Lights case, this happened in my home state of Arizona. In fact, it occurred in 1975, which is when my wife's father moved to the White Mountains to a town called Show Low. This is near the place Travis Walton's experience occurred, so I am quite familiar with the area. I have visited many times over the past 40 years. So, this is one reason it caught my attention. However, the case itself also has intrigued me.

I also must let you know that I have met Travis on a number of occasions, we are acquainted, and, in fact, he gave me an autographed copy of his book. He is a down to earth man, whom I believe to be honest, forthright, and truthful regarding his abduction. He is not sensationalistic. To the contrary, he is shy and reserved. However, I have seen him boldly shut down people who try to embellish his account in any way. His story has remained the same, which I consider a factor in his credibility. Much of this account is taken from the book, *Fire in the Sky* by Travis Walton himself.

In 1975, Travis and six fellow loggers were clearing a forest area in the Apache-Sitgreaves National forest located in the White Mountains of Arizona under a contract with U.S. Forest Service. Arizona has the largest stand of Ponderosa Pines in the nation. The forest can develop thick underbrush from lots of water gained from heavy snows in the winter, to heavy rains during the monsoon season. This results in the need to clear the forest, since summertime temperatures create quite a fire hazard. In fact, some years ago, more than a half million acres were burned in the Rodeo Fire. It was a very sad thing to see. So, this was important work they were doing. Travis and his crew were residents of a small community known as Snowflake, AZ. It is an area with a strong Mormon population, with logging and ranching as economic mainstays. So, if you have ever lived in a small community, you are acquainted with the fact that everybody knows everybody and their business too. This is a factor in the abduction case as we will discuss later.

The crew included Travis Walton, crew leader Mike Rodgers, Dwayne Smith, Kenneth Peterson, Steve Pierce, John Goulette, and Allen Dalis. After a long hard day of manhandling chain saws, they began packing up their equipment at about 6:00 p.m. This was close to sundown on that November day. As they headed down the bumpy logging road in their 1965 International Harvester crew cab pickup, they saw a light in the forest a few hundred yards down the road. They didn't know what it was, but surmised it was a camp fire, downed airplane, even just the sunset. Boy, were they wrong!

As they approached, they saw a "flying saucer!" Awestruck, the crew leader, Mike Rodgers, stopped the truck a few dozen yards from the object. The craft was a classic flying saucer about 20 feet in diameter, hovering about 15 feet above the ground. It was emitting a yellow glow and was completely silent.

At this point, the stunned observers began to react. One shouted, "Is this really happening?" Another crouched behind the truck in abject fear. And one, Travis Walton, inexplicably left the truck and slowly walked toward the craft to get a better view. All the while, his crew mates screaming, "Get back in here you crazy son of a bitch!". Crazy indeed, because he was about to pay dearly for that impetuousness.

Travis, who was now only a few feet away from the craft, heard high and low- pitched mechanical sounds and high-pitched beeping sounds amid a low rumbling noise like that of machinery. Then, the tones changed, and the craft started to wobble. He was struck by a bluish beam of light that shot from the bottom of the craft. The crew was terrified to see that beam strike Travis and knock him back a good ten feet, where he landed in a lifeless pile, thrown like a rag doll. They were certain he was killed. Complete panic now gripped the crew, and Mike Rodgers responded to the screaming pleas of the crew to "Get the hell out of here!"

After a few minutes of fearful flight down a bumpy and hazardous logging road, in the dark, in a truck with 1965 technology, they skidded to a stop to ponder what they should do next. After some heated debate fueled by fear and adrenaline, Mike Rodgers insisted he was going back for Travis, and anyone who didn't want to go could wait right here in the dark forest.

So, back they all went. However, rather than finding Travis where they last saw his lifeless body, they found no Travis and no saucer! They then made a report to the Navajo County Sheriff's Office. After hearing the fantastic story from the obviously terrified crew, the Sheriff's Office organized a search party and would deal with the "UFO stuff" later.

Happily, Travis did reappear five days later. But, this caused one of the biggest searches in the history of Arizona, created divisions within families and the community, and resulted in murder accusations and polygraph tests of the seven crew members.

I am going to leave the rest of the story for you to discover on your own. But suffice it to say, this was a case study in the reaction of community, law enforcement, experiencers, debunkers, media, and others that has been unrivaled since. This is such an important case. I highly recommend you research it.

The Phoenix Lights 1997

The Phoenix Lights were a series of unidentified flying objects observed in the skies over the states of Arizona and Nevada on the night of March 13, 1997.

This case was responsible for my serious interest in UAP phenomena. Not because I saw the craft, which I did not, but because it happened in my hometown. There was a lot of news coverage, people were talking about it, and it intrigued me. It brought home the reality of UAP.

Thousands of people made reports to MUFON, NUFORC, local police, the military, and media in what was later mapped to be a nearly straight line from Henderson, Nevada through Tucson, Arizona. In fact, the objects flew over the city of Phoenix—a metropolitan area of nearly five million people. The entire affair occurred from about 7:00 p.m. through 11:00 p.m.

There were technically two events involved. The first was a boomerang-shaped formation of lights seen passing over the state. This was the sighting most people reported, but only one video is known to exist. The second, around 10:00 pm was a display of stationary lights over the Estrella Mountains on the southern border of the metro-Phoenix area. This was seen by thousands of people and was recorded on video by multiple sources.

The United States Air Force claimed the second group of lights was flares dropped by aircraft that were on training exercises from an Air National Guard unit out of Maryland.

At first, there was minimal news coverage at the time of the incident. But, on June 18, 1997, USA Today ran a front-page story that brought national attention to the case. This was followed by news coverage on multiple national television networks. The case has gained world-wide notoriety and is thought to be one of the most important cases in the modern UFO era. This case has appeared on UFO-related documentary television, including specials produced by the History Channel, the Discovery Channel, and an excellent book and documentary by Dr. Lynne Kitei. I highly recommend these resources for your more in-depth research of this case. There is much more to this case than is generally talked about. Moreover, Dr. Kitei puts on an event each year on the anniversary of the sighting that is also excellent. If you have the means, I encourage you to attend.

It's worth noting that Arizona governor Fife Symington held a news conference, stating that, "they found who was responsible." He proceeded to ridicule the situation by bringing his aide on stage dressed in a cheesy alien costume to the laughter of the media and those present at the televised briefing. However, later in an interview by James Fox, for his documentary *Out of the Blue,* Symington said that he had witnessed one of the "crafts of unknown origin" during the 1997 event. *"A lot of people saw it, and I saw it too."* This nearly floored James Fox when the Governor shyly slipped in his admission since he had ridiculed the sighting before. The governor indicated he ridiculed this because he wanted to make light of the situation to reduce the hysteria that had gripped the city. He was already having a hard time governing, and this admission would have ruined all his credibility. It was largely a political tactic.

He further stated, "I'm a pilot and I know just about every machine that flies. It was bigger than anything that I've ever seen. It remains a great mystery. Other people saw it, responsible people. I don't know why people would ridicule it." He continued, "As a pilot and a former Air Force Officer, I can definitively say that this craft did not resemble any manmade object I'd ever seen. And it was certainly not high-altitude flares because flares don't fly in a formation." Symington said that he contacted the military asking what the lights were. The response was, "no comment." He pointed out that he was the governor of Arizona at the time, not just some ordinary civilian, and still got stonewalled!

Frances Emma Barwood, a Phoenix city councilwoman who launched an investigation into the event at that time, said that of the more than 700 witnesses she interviewed, "The government never interviewed even one." Her interest and suggested investigation to the Phoenix City Council got her mocked, and she eventually lost her council seat because of it.

This is a case study on so many levels. The UAP itself, the reaction of the public, police, military, media, and the political ramifications of UAP on a society.

This is a must research case if there ever was one!

Wright Patterson Airforce Base

Just a quick reference here to the famous air force base

located in Dayton, Ohio. My good buddy Ray Szymanski had a long career there, and he recounts with considerable wit his experiences there in his book *Fifty Shades of Greys*. Also, it is a well-known story that Senator Barry Goldwater asked General Curtis LeMay to see the "Blue" room where they allegedly hold UFO debris at the base and was roundly denied. In fact, LeMay cursed at Goldwater and said, "Don't ever ask me that again!" I mention this as a classic case more for the point that the UFO secrecy is no fiction. It's a fun case to research, so I recommend it.

So, you may notice each of these cases has a different form of evidence that makes them compelling. Some are strong on physical evidence and a thorough investigation, such as the Soccorro case. The Phoenix Lights case contains video and photo evidence as well as eyewitness testimony from many observers. The Los Angeles case provides historical documentation along with photographs and witness testimony. The Roswell case has some of everything. And the Washington, DC case has strong radar and eyewitness evidence as well as documentation. And, of course, the Travis Walton case is truly unique in all of its forms of evidence, uncommon in abduction cases.

You may discover even more in your own research of these classic cases and will no doubt develop a sense for the kind of evidence that speaks to you the most.

Note to Self
Each case has strengths and weaknesses when it comes to evidence.

Chapter 7

Ufology: Separating the Wheat from the Chaff

Ufology, the study of Unidentified Flying Objects (UAP) is a world unto itself. Like any topic of study, you have disagreement among experts. There are many reasons for this, such as differing data, differing opinion of what the data suggests, personal opinions, the influence of political forces, and the list goes on.

So, the topic contains "wheat-like" good information and "chaff-like" bad information. And the bad certainly outweighs the good. It's like looking for gold. You must dig through a lot of worthless ore to get to the gold nuggets. But, correctly researching UAP, just like digging for gold, it is worth the effort. One just needs to develop the skills to separate the wheat from the chaff. So, let's take a look at some of the chaff-like information you have to dig through.

For example, there is the intentional **misrepresentation** of the data. This could be due to the desire to generate income from click bait presentations of either distortion of ambiguous data or outright false information. Again, motivated by greed, politics, or a desire for fame. People putting this out there as "real" have no genuine interest in proving the reality of UAP based on research and facts. They just want to benefit from it in some way. There is also misinformation due to inadequate investigation and research, or even due to outright gullibility.

There are proven cases of **disinformation** by government,

military, and other officials designed to confuse, mislead, or in some other way interfere with the research, study, and presentation of legitimate UFO evidence. There are famous cases such as the Paul Benewitz case, the USAF Roswell Report, and others that were government "caught red handed lying" cases.

Proof of this disinformation method is aptly admitted to in the 1953 CIA's so-called "Robertson Panel" memo that clearly indicated disinformation is a key technique utilized by the government to delegitimize UAP research. The "Condon Report" prepared by the University of Colorado's Edward Condon was an obvious and embarrassing whitewash of the UAP results from "Project Blue Book" that was used to justify closing Project Blue Book.

There are also other agendas at play. There are the ever-present **debunkers** who seek to discredit people and information in ufology. This may stem from a fear that there may actually be "others," which upsets their world view, and/or scientific or religious belief systems. They may be genuinely offended by the topic. They may have a career to protect. They may just get off on looking like the smartest guy in the room. They may be "wanna-be-scientists" who shoot from the hip on social media misapplying scientific terms and ideas to look smart. Or they may be assigned to obfuscate the UAP community by intelligence agencies.

Whatever their motivations may be, I consider them the other side of the coin with true believers. I say this because debunkers will dismiss, at all costs, evidence and logic to debunk a case—just as a true believer may ignore data and facts to support their opinion in the reality of UFOs. Debunkers are often associates of government intelligence agencies in some way, who are incentivized to discredit UAP. They also may be scientists who make their living from grants given by those agencies. They may be academics who fear for their credibility and tenure at a university. They may even be jerks who simply get off on sowing seeds of discontent among a community, which was highlighted in an earlier chapter on hoaxes.

There are also the ever-present **"True Believers"** who must prove that "Others" exist to support their world view or other belief systems. So, a word of caution regarding judgment of people displaying a strong belief.

I greatly respect the opinions and feelings of every sincere person interested in this topic whether it be a believer or disbeliever.

I do so for various reasons such as: Who they hell am I to judge another person based on their belief? Some people, like me, do not necessarily "believe" others exist because they have seen no proof, but do hold out that it is "possible" they exist. And I hold this view because of the evidence I have seen based on my personal research. UAP positively exist. The data proves it. However, the data has not yet proven "who" is behind UAP.

While I have had no personal experience with a UFO or an encounter with an "other" intelligent life form, if I did, I bet I would be not just a "believer," but a "knower!"

So, some people base their belief on their personal experience. There is nothing like seeing a UFO in-person or having an encounter with a so-called "other" to make one a believer. I certainly appreciate the impact that would have on a person's belief system. Since I was not there, I start with the person's sincere belief in what they saw or experienced. I am not going to dismiss that without a thorough investigation. Even if an investigation does not yield proof of that sighting or experience, my opinion of the evidence is just that, an opinion. I hold that their experience and viewpoint that it was a genuine ET experience is as valid as any conclusion I may make. Again, I wasn't there and I don't "know" that it wasn't an ET encounter. Frankly, the source of UAP phenomena has not yet been publicly identified or proven by anyone to our knowledge.

There are other believers that develop this view based on religious or spiritual beliefs. Some conflate "New Age" thinking with the paranormal, UFOs, and other topics to satisfy their spiritual needs. Again, I am no one's judge, but I do reserve the right to formulate an opinion based on the actual evidence.

So, how does one sift through all the noise and nonsense of

the "Chaff" of misrepresentation, disinformation, belief versus research, investigation, and analysis to determine what is the "Wheat" when it comes to UAP information?

Let's talk about the **misinformation** crowd. This is the largest group purveying chaff in the field of ufology, and it is largely because of the internet.

UAP is a visual-phenomena, and the internet is the largest, most effective disseminator of photographic and video information on the face of this planet! It is the perfect medium for making visual information available to the masses. It is highly effective and is a wonderful source to see things we may never see in our lifetime without it.

I marvel at the massive amount of videos and photographs posted from all over the world. Just reflect on how many videos of amusing pets, wild animals, beautiful forests, cities, wonders of every kind show up on websites, social media, advertisements, and every other conceivable forum. It is astounding!

Everyone wants to see actual UFO/UAP videos or photos. Everyone is hoping for proof! This would be of immense interest to scientists, religionists, governments, artists, engineers, aviation industry, military, advertisers, the stock market, literally everyone on earth. This fact, unfortunately, makes us easy pickings for opportunistic hucksters to use that natural curiosity and interest to enrich themselves financially or with fame.

So, how do you distinguish legitimate information from illegitimate information? Just like with everything else, you need to think critically. Do not take anything at face value. Question everything and everyone. Be skeptical. Assume falsehood, not truth, and work back from there. Prove it or lose it, that's my motto!

So, when someone says "I know" this or that, your BS meter should be pegging! The fact is no one knows anything! Anything and everything in ufology is unproven. Don't get me wrong, there is a ton of good evidence that suggests an unexplained phenomenon exists. It is this evidence that you should be interested in. You should be researching and vetting this information and those who present it. It's just like Stanton Friedman always said: "85% of UFOs are identifiable, I don't care about those. I care about the 15% you can't identify!".

So, avoid the click bait, the sensationalistic, even the many websites and YouTube videos, or other social media photos and videos that are meant to entice or entertain. This is not serious research and only clouds and diminishes the real and credible information.

In an earlier chapter, I related extensively what good evidence looks like and how to identify it. As for misidentification, it is as much about the motive of the purveyor as it is the misleading evidence.

Now the **"disinformation"** crowd is a bit trickier to deal with. First, you must become aware of the fact that there are people and organizations who knowingly put forth false information (disinformation). Yes, you as an interested person of the UAP phenomenon are a target of disinformation agents. They want to affect your thinking and viewpoint of this topic. They want to control the message and are not afraid to use propaganda, twisted truths, outright lies, ridicule, reputation assassination, you name it, they'll use it. Are you surprised and intimidated by this? Don't be. You can identify this behavior and avoid it. And believe me, avoid engaging with individuals who are professional debunkers, disinformation agents, or obnoxious contrarians who are only interested in saying "black" to your "white." It's a waste of time and a distraction and irritation. Who needs that crap? You have a wonderful mystery to explore. Don't let these jerks rob you of the joy and satisfaction in learning about this most interesting of topics!

Finally, a word about UAP and pop culture. This is both a good thing and a bad thing. As with any public interest topic, there will be people who take it seriously and those who do not. But the very fact that UAP has become a pop cultural item is likely a good thing. After 70 plus years of ridicule, the public is now taking the topic much more seriously in many respects. When something becomes popular in our culture, there are more eyes on it. More people digging deeper and revealing facts and falsehoods. However, the converse is also true. The more people interested in it on a pop culture level see it as entertainment, a toy, and not taken seriously. So, balance is needed. You may be surprised at the following example I am going to relate when it comes to pop culture and ufology.

Take for example the Kardashian family. This is a family who has parlayed their wealth, famous associations, beauty, and other influences into a pop culture icon in itself. Many say they are famous for nothing. Many people don't take them seriously for their shameless pursuit of fame and fortune based on no apparent accomplishments, other than those I just listed. However, there are many people who take them seriously. They have skillfully created a great deal of wealth, power, and influence through their television programs, clothing lines, and beauty products, earning tens of millions of dollars. Ridicule them if you wish, but they have also made successful use of pop culture.

So, when it comes to UAP and pop culture, you will see that there are in fact two sides to it. There is the very serious scientific view, and then there is the less serious, fun to imagine point of view of UAP in pop culture.

One refrain that often comes from many on social media and by the debunker crowd is the criticism of money in ufology. The Bible is often misquoted when it states that "money is the root of all evil." The quote is actually "the love of money is the source of many injurious things". Do you see the difference? Money is not evil. The "loving of it" or improperly placing its value above all other things is wrong.

In order to accomplish literally anything of value in life takes money. Think about it in everyday life. We need money to live. Food, clothing, shelter, transportation, fuel, absolutely every necessity and anything we merely want has a cost. Right now, I am spending money to write this book. Right now, you are spending money to read this book.

If a person is engaging in legitimate and sincere effort to research the UAP topic, why is it some offense against humanity to earn money to do so? This is ridiculous and another technique to undermine or otherwise discredit someone's sincere effort and expense, and is frankly the product of jealousy, self-loathing, or a small mind. Of course, there are also examples of people who obviously exploit the topic for profit with no real conviction for the truth. That is despicable and shameful.

So, remember to have your "bull-shit" meter calibrated so you can identify and avoid the chaff, while also correctly identifying the wheat!!

<u>Note to Self</u>
There is nothing like seeing a UFO to make one a believer!

Chapter 8

The Cover-up is Real

Whether it is a family, a corporation, or a government, organizations have issues they wish to keep private. This is based on normal human logic and behavior. It is not necessarily a conspiracy, which is defined as "a secret plan by a group to do something unlawful or harmful." Creating and implementing a plan to accomplish goals for the organization, be it a family, group, or formal organization, is, in fact, not a conspiracy. It is a necessary step in managing said organization and is not intended to be devious, unlawful, or harmful.

It is important to have a correct understanding of this definition to intelligently discuss the cover-up of UAP information held by governments, private companies or individuals.

A "cover-up" is generally thought to be an attempt to hide information to prevent knowledge of this information by others. Likely, the purpose is to hide a failure, to protect from perceived harm, or to control others. This can stem from a positive or negative motivation.

I wanted to make this distinction because the topic of governmental cover up can be interpreted as a conspiracy, which is a rabbit hole I want to avoid at this point in your research efforts. You need to be aware that conspiracies do exist, but you don't want to cloud your research with them until you gain some experience. The fact is that a cover-up can be both legitimate and illegitimate based on the sensitivity of the information and motivation of the coverer.

Cover-up is a less deep rabbit hole than conspiracy, and is enough on your plate at this point of your research career. I just want to make you aware of the concept of cover-up and its role in UAP research. If conspiracy appeals to you, then I just recommend you go there only after gaining some experience to help you stay grounded.

There is no doubt the United States government, and other governments of this planet, do cover up the information they have accumulated regarding UAP for decades. There are legitimate reasons for not sharing information or "covering it up."

We are all quite aware that our government has classifications of information that range from public, sensitive, secret, to even top secret, with even more classifications above and in between. This is not nefarious, but a logical and necessary way of managing information.

Let's look at a few examples to illustrate. Think about it, we classify information and access to it even in our families. Do parents discuss intricacies of their finances with the children? For example: "Hey little Joey, I got 4% interest on our new car loan! Oh, cool dad, good negotiation skills you have there!" Or, do we discuss our marital problems on the internet? "Hey, Facebook friends, my husband just made a really poor decision, what do you think?" Do you share how much money you make with others? No, those things are basically not anyone else's business and for good reason. Not a conspiracy or nefarious, just the use of good judgement to keep private things private so as not to do any harm to yourself or the people you love.

Now, there are forms of information that are important to share like, "Hi neighbors. The electric company said our power will be back on by noon." Or, "Don't use the backroad. It's flooded." We may feel compelled to share some information like, "I know a good contractor for your project," or, "That mechanic was not good or trustworthy." Again, examples of reasonable sharing of some information and reasonable withholding of other information.

On a corporate level, should all information be shared? Of course not. There are issues of proprietary information that, if shared, might result in a loss of competitive edge or market share. There may be confidential information regarding customers' personally identifying information such as social security numbers, or employee medical information. It is critically important that this information is protected, or "covered up" as it were. Conversely, some information should be shared, such as stock performance of a publicly held company, new product releases, or even negative information such as an investigation of fraud. Though unflattering, legal and ethical reasons require the need for transparency. There is also a need for transparency with governments and UAP information.

However, there are legitimate reasons why some of this information may be covered up. For example, intelligence agencies need to protect the identity and methods of intelligence gathering techniques, sources, and personnel. If a UAP report involved these facets of information, then covering up that information would be legitimate. So, an appropriate response would be to make the report available, but redact sensitive or secret information to protect these methods and individuals from those who may do them harm. What would be illegitimate would be to cover up the report altogether. So, I hope I have illustrated the need to properly classify and manage information on many levels.

Now, let's be specific about the cover-up of UAP information in the United States since the beginning of the modern UFO phenomenon. I make that distinction because, since the 1940s, it is publicly known that the US government has acknowledged the phenomenon, studied it, and covered up much of the results of that study. Even though the phenomenon had been around since ancient times, there is more information available in the form of written record during our modern times to investigate, so let's focus on that.

American government agencies such as the FBI, CIA, DIA, all military branches, and many others not listed here, have produced documentation of their UFO related activities. So, there is certainly an abundance of UFO reports compiled in the open and official UFO research and investigation programs such as projects SIGN, GRUDGE, and BLUE BOOK. There has been a recent revelation of a more recent UAP investigation program called the Advanced Aerospace Threat Identification Program (AATIP), which was officially funded from 2007 through 2012 for about twenty-two million dollars.

While none of these programs were secret in their entirety, some elements of them clearly were kept secret. Information of higher value was handled in a different, confidential path than the more mundane cases that were made public. Be that as it may, a great deal of information was shared, but a great deal more was not.

I highly recommend you read the book, *The FBI-CIA-UFO Connection*, by Dr. Bruce Maccabee. It is one of the most interesting, detailed, and telling documentations of involvement of these agencies and the UFO phenomenon. The documentation presented in this publication alone is proof of disclosure in my mind, but that's for another chapter.

The US government has a long, documented, and provable interest and participation in UAP phenomena despite their claims. As mentioned, the modern UFO phenomenon caught the attention of the US government in 1947. After the Maury Island incident in Washington State, followed by the Kenneth Arnold sighting in June of that year, and, of course, the Roswell incident in July, the government had to respond to the phenomenon and the excitement it created in the American public consciousness.

Amid many other sightings and reports, in 1948, Project SIGN was set up to formally study the phenomenon from a military, intelligence, and scientific perspective. UFOs were rapidly becoming the topic of newspaper, magazine, and other media. The public was reaching a state of hysteria. This was putting pressure on the government to act. However, the primary concern had to be whether this represented a threat to the nation or not.

This is the study that resulted from the famous memo by Lt. General Nathan Twining of the Air Materiel Command at Wright Field who stated that "the phenomenon reported is something real and not visionary or fictitious." Regarding "cover-up," this project was not made public until years later when referenced by Edward

J. Ruppelt in his book titled, *The Report on Unidentified Flying Objects* in 1956. This is also the project that allegedly produce the "The Estimate of the Situation" memo that favored the ET hypothesis, but was roundly rejected by the Chief of Staff General Hoyt Vandenburg. The memo was reportedly destroyed, and no copies have ever surfaced.

As you can tell from the new project name, Project GRUDGE, the motive of UFO study changed from identification to suppression of UFO reports. In fact, this institutionalized debunking by the government, and was the mother of all cover-up activity since that time.

Though there were many who opposed UFO study within the military and intelligence communities, there were also many who supported continued legitimate study. The public was seeing a great many things and deserved to have its government make an earnest effort to acknowledge and study the phenomenon. As a result, Project GRUDGE ultimately concluded and was replaced with Project BLUE BOOK.

Project BLUE BOOK was thought to be one of the more productive government studies of UAP, even though there was constant pressure to minimize or debunk UFO reports submitted by citizens from around the country. However, many reports came from the military, scientists, law enforcement, and other credible people. This could not simply be ignored. So, some 12,618 reports were investigated, resulting in 701 truly unexplainable sightings or encounters.

Entire books have been written about project BLUE BOOK, but just know that it is a real treasure of UAP information. Some of its conclusions are clearly debunker, obfuscation, and insulting to one's intelligence. However, some of the cases are very compelling. I encourage you to read *UFOs and the National Security State* Vols 1 & 2 in which Richard Dolan provides analysis of many of these reports. This is one of the most comprehensive and cogent consideration of the project I have ever read.

Now, since Project BLUE BOOK was terminated in 1969, we were told by our government that they were no longer interested in, nor did they study, the UFO phenomenon, as it has no scientific or military significance. This was the conclusion of Edward U. Condon, who led the University of Colorado report commissioned by the USAF in 1968. Of course, this is utter BS. The US government continued to be interested and continued studying it all along. They simply moved the classification from public to secret to get it out of the public's eye.

Since the closure of BLUE BOOK, ufologists have insisted that the government was continuing its UFO programs under so called "BLACK projects." These are projects that are considered secret and not subject to public or even congressional scrutiny. They are on a "need-to-know basis" only, compartmentalized, and kept from many within and without government.

Well, in October of 2017, the lid to this secrecy was blown off the pot! It was revealed that the United States Government had indeed a program to study UAP phenomena. A group of former military, intelligence, scientific, and political individuals formed a private company named To The Stars Academy of Arts and Science designed to release UAP information gathered and analyzed by a government program titled Advanced Aerospace Threat Identification Program or AATIP.

This program was created by three United States senators and funded with a twenty-two million dollar appropriation within the Department of Defense's Defense Intelligence Agency for the years of 2007 through 2012. Though the formal program funding has ended, the threat assessment continues to occur to this day. Likely, this activity has continued to be performed under a department's operating budget rather than funded by a stand-alone appropriation.

So again, we have proof of the government's cover up of its interest and involvement in the research of UAP. Literally since Project Grudge in 1949, the government has continued to cover up, minimize, debunk, and ridicule the UAP phenomenon. WHY? If there is nothing to it, why continue to study it, why continue to ridicule it?

I have always thought to myself: When does someone stop

asking a question? When they get an answer . . .

Cover-up, both for legitimate and illegitimate reasons, is a proven and real aspect of UAP phenomena. It is not a crazy conspiracy theory. It is very, very real.

Nonetheless, this is a fascinating part of UAP history, and is an excellent source of material for you to research and study. So, go forth, research, study, and develop your own thoughts and ideas and enjoy this wonderful mystery!

Note to Self
When does someone stop asking a question? When they get an answer…

Chapter 9

Political Science and Ufology

I hate to do this to you again, but it's back to class. Political Science to be exact!

There is a lot of discussion today on the topics of cover-up, disclosure, alien agenda, cattle mutilations, crop circles, consciousness, dimensional and universal theory, and other ways to interpret UAP phenomena. We yearn to know what it is and why it is happening. Curiously, human logic demands we fill in the blanks of our understanding. Why so? It is a reflection of the political science of human cognition. So, we need to examine for a moment why we think the way we do about UAP.

One of the most fascinating papers I have ever read is titled, "Sovereignty and the UFO" by Alexander Wendt and Raymond Duvall. I want to quote and comment extensively from this paper to help you understand the context by which you and I are curious and pursuing the question of UAP. It is a fascinating consideration of our motivation in seeking this truth and how it may affect your pursuit of this subject. It will aid you in an objective basis for your opinions and beliefs as you wade through this topic. I think it is critical to understand this as you research the UAP topic.

Most of all, it will be especially helpful in understanding the history of UFOs and our government's schizophrenic political response to the phenomenon. This helps us to understand the incredible pushback to researching the UAP subject by the majority of society. So, this school of thought starts with one of the psychological underpinnings of the subject termed "anthropocentrism."

So, what is anthropocentrism? According to Webster it is defined as: "Interpreting or regarding the world in terms of human values and experiences." We see and interpret everything from our point of view. This is logical, but can be limiting, and accounts for some of our myopathy. Let's see how Wendt and Duvall explain how we and our government apply anthropocentrism to the UAP topic and the notion of "sovereignty, or the right to rule."

> "Modern sovereignty is anthropocentric, constituted and organized by reference to human beings alone. Although a metaphysical assumption, anthropocentrism is of immense practical import, enabling modern states to command loyalty and resources from their subjects in pursuit of political projects. It has limits, however, which are brought clearly into view by the authoritative taboo on taking UFOs seriously. UFOs have never been systematically investigated by science or the state, because it is assumed to be known that none are extraterrestrial. Yet in fact this is not known, which makes the UFO taboo puzzling given the ET possibility."

So, governments consider UAP as a potential threat to their "sovereignty" because the answers to the question of what UAP is may reveal a power greater than the state. UFOs are considered an existential threat to any government. To them, that notion is as serious as a heart attack! Notice how intense this view is as expressed by Wendt and Duvall:

> "Few ideas today are as contested as sovereignty, in theory or in practice. In sovereignty theory scholars disagree about almost everything—what sovereignty is and where it resides, how it relates to law, whether it is divisible, how its subjects and

objects are constituted, and whether it is being transformed in late modernity. Sovereignty is the province of humans alone. Animals and Nature are assumed to lack the cognitive capacity and/or subjectivity to be sovereign; and while God might have ultimate sovereignty, even most religious fundamentalists grant that it is not exercised directly in the temporal world."

Wow! When sovereignty is contested today, it is only among humans, rather than with animals, Nature or God. In this way, modern sovereignty is anthropocentric and organized by reference to human beings alone. So, what if there is a power other than humans (i.e. aliens) that could rule? Notice their consideration of that potential:

"Anthropocentric sovereignty might seem necessary; after all, who else, besides humans, might rule? For millennia Nature and the gods were thought to have causal powers and subjectivities that enabled them to share sovereignty with humans. Authoritative belief in non-human sovereignties was given up only after long and bitter struggle about the "borders of the social world," in which who/what could be sovereign depends on who/what should be included in society."

So, first and foremost, governments are motivated to protect their sovereignty. And they see UAP, or its intelligent "Others," as a threat to their existence. Is that understandable? Notice Wendt and Duvall's logic here:

"It is in anthropocentric terms that humans today understand their place in the physical world. Suggestions of animal consciousness fuel calls for animal rights, for example, and advocates of "Intelligent Design" think God is necessary to explain Nature's complexity. Yet, such challenges do not threaten the principle of sovereignty, the capacity to decide the norm and exception to it, must necessarily be human.

Animals or Nature might deserve rights, but humans will decide that; and even Intelligent Designers do not claim that God exercises temporal sovereignty. With respect to sovereignty, at least, anthropocentrism is taken to be common sense, even in political theory, where it is rarely problematized."

Now, this is where it starts to get really scary:

"This 'common sense' is nevertheless of immense practical significance in the mobilization of power and violence for political projects. Modern systems of rule are able to command exceptional loyalty and resources from their subjects on the shared assumption that the only potential sovereigns are human.

Imagine a counterfactual world in which God visibly materialized (as in the Christians' 'Second Coming,' for example): to whom would people give their loyalty, and could states in their present form survive were such a question politically salient? Anything that challenged anthropocentric sovereignty, it seems, would challenge the foundations of modern rule."

So, why do governments deny and obfuscate the UAP reality? It would potentially change the entire nation-state system under which this world is currently ruled. Like we said, to them, it is existential, as serious as a heart attack!

In this paper, Wendt and Duvall divide the topic into four sections: They describe the "UFO *taboo*" in order to set the empirical basis for theoretical intervention. In the next section, they make an immanent critique of the authoritative *claim* that UFOs are not extraterrestrial (ET). They perform a theoretical analysis of the metaphysical *threat* that the UFO poses to anthropocentric sovereignty. They conclude with some *implications* for theory and practice.

First, in describing the *taboo* on UAP discussion or study, they use the 1990 Belgium UAP incident as an example of the governments' dismissal of good case data. Notice the account and conclusions they make:

> "On March 30-31, 1990, two Belgian F-16s were scrambled to intercept a large, unidentified object in the night sky over Brussels, which had been observed by a policeman and ground-based radars. The pilots confirmed the target on their radars (never visually) and achieved radar lock three times, but each time it responded with violent turns and altitude changes, later
> estimated to have imposed gravitational forces of 40gs. In a rare public statement the Belgian defense minister said he could not explain the incident, which remains unexplained today.
>
> One might expect unexplained incidents in NATO airspace to concern the authorities, particularly given that since 1947 over 100,000 UFOs have been reported worldwide, many by militaries. However, neither the scientific community nor states have made serious efforts to identify them, the vast majority remaining completely uninvestigated. The science of UFOs is minuscule

and deeply marginalized. Although many scientists think privately that UFOs deserve study, there are no opportunities or incentives to do it. With almost no meaningful variation, states—all 190+ of them—have been notably uninterested as well. A few have gone through the motions of studying individual cases, but with even fewer exceptions these inquiries have been neither objective nor systematic, and no state has actually looked for UFOs to discover larger patterns. For both science and the state, it seems, the UFO is not an 'object' at all, but a non-object, something not just unidentified but unseen and thus ignored.

The authoritative disregard of UFOs goes further, however, to active denial of their object status. Ufology is decried as a pseudo-science that threatens the foundations of scientific authority,14 and the few scientists who have taken a public interest in UFOs have done so at considerable cost. For their part, states have actively dismissed 'belief' in UFOs as irrational (as in, 'do you believe in UFOs?'), while maintaining considerable secrecy about their own reports.

This leading role of the state distinguishes UFOs from other anomalies, scientific resistance to which is typically explained sociologically. UFO denial appears to be as much political as sociological— more like Galileo's ideas were political for the Catholic Church than like the once ridiculed theory of continental drift. In short, considerable work goes into ignoring UFOs, constituting them as objects only of ridicule and scorn. To that extent one may speak of a 'UFO taboo,' a prohibition in the authoritative public sphere on taking UFOs seriously, or 'thou shalt not try very hard to find out what UFOs are.'

"Still, for modern elites it is unnecessary to study UFOs, because they are known to have conventional—i.e., non-ET—explanations, whether hoaxes, rare atmospheric phenomena, instrument malfunction, witness mistakes, or secret government technologies. Members of the general public might believe that UFOs are ETs, but authoritatively We know they are not. In the next section we challenge this claim to knowledge. Not by arguing that UFOs are ETs, since we have no idea what UFOs are—which are, after all, unidentified. But that is precisely the point. Scientifically, human beings do not know that all UFOs have conventional explanations, but instead remain ignorant."

Secondly, notice how they expose the illogical *claim* that the governments know UFOs are not extraterrestrial!

"If any UFOs were discovered to be ETs it would be one of the most important events in human history, making it rational to investigate even a remote possibility. It was just such reasoning that led the U.S. government to fund the Search for Extra-Terrestrial Intelligence (SETI), which looks for signs of life around distant stars. With no evidence whatsoever for such life, why not study UFOs, which are close by and leave evidence?

States seem eager to 'securitize' all manner of threats to their societies or their rule. Securitization often enables the expansion of state power; why not then securitize UFOs, which offer unprecedented possibilities in this respect? And finally, there is simple scientific curiosity: why not study UFOs, just like human beings study everything else? At least something interesting might be learned about Nature. Notwithstanding these compelling reasons to identify UFOs, however, modern authorities have not seriously tried to do so.

This suggests that UFO ignorance is not simply a gap in our knowledge, like the cure for cancer, but something actively reproduced by taboo. Thus, our puzzle is not the familiar question of ufology, 'What are UFOs?' but, 'Why are they dismissed by the authorities?' Why is human ignorance not only unacknowledged, but so emphatically denied? In short, why a taboo? These are questions of social rather than physical science, and do not presuppose that any UFOs are ETs. Only that they might be.

The UFO compels decision because it exceeds modern governmentality, but we argue that the decision cannot be made. The reason is that modern decision presupposes anthropocentrism, which is threatened metaphysically by the possibility that UFOs might be ETs. As such, genuine UFO ignorance cannot be acknowledged without calling modern sovereignty itself into question. This puts the problem of normalizing the UFO back onto governmentality, where it can be 'known' only without trying to find out what it is—through a taboo. The UFO, in short, is a previously unacknowledged site of contestation in an ongoing historical project to constitute sovereignty in anthropocentric terms. Importantly, our argument here is structural rather than agentic.

> We are not saying the authorities are hiding The Truth about UFOs, much less that it is ET. We are saying they cannot ask the question."

Thirdly, the governments cannot admit the reality of intelligent life here on our planet without abdicating their authority according to this view. It is no wonder we have been stonewalled at every turn when investigating UAP!

How about the viewpoint of science and religion as institutions? Would UAP pose a *threat* to them as well?

> "In the UFO context such anti-realism is problematic, since its political effect is ironically to reinforce the skeptical orthodoxy: if UFOs cannot be known scientifically then why bother study them? As realist institutions, science and the modern state do not concern themselves with what cannot be known scientifically. For example, whatever their religious beliefs, social scientists always study religion as 'methodological atheists,' assuming that God plays no causal role in the material world. Anything else would be considered irrational today. By not allowing that UFOs might be knowable scientifically implicitly embraces a kind of methodological atheism about UFOs, which as with God shifts attention to human representations of the UFO, not its reality.
>
> Yet UFOs are different than God in one key respect: many leave physical traces on radar and film, which suggests they are natural rather than supernatural phenomena and thus amenable in principle to scientific investigation. Upon close inspection many UFOs do turn out to have conventional explanations, but there is a hard core of cases, perhaps 25 to 30 percent, that

seem to resist such explanations, and their reality may indeed be humanly unknowable—although without systematic inquiry we cannot say.

Nevertheless, in the context of natural phenomena like UFOs
agnosticism can itself become dogma if not put to the test, which requires adopting a realist stance at least instrumentally or 'strategically,' and seeing what happens. This justifies acting as if the UFO is knowable, while recognizing that it might ultimately exceed human grasp."

Fourthly, their argument is that UFO ignorance is political rather than scientific. What are the *implications* of this? Science derives its authority from its claim to discover. Since governments claim these facts include that UFOs are not ETs, we have to show that this fact is not actually scientific.

So now, we can conclude there is no *scientific* reason to *ignore* the study of UAP. We considered the government's arguments for UFO skepticism and show that none justifies rejection of the Extra-Terrestrial Hypothesis (ETH).

It is not known, scientifically, that UFOs are not ETs. And, to reject the ETH may be rejecting a true explanation. Of course, this does not mean that UFOs are ETs either, but it shifts the burden of proof onto skeptics!

As a new student of the UAP topic, you will likely pick up on key arguments utilized by debunkers (remember, skeptics are not the same as debunkers) and we will touch on a few and the observations Wendt and Duvall made from a political science perspective.

The first claim of debunkers is that there is "no evidence" to support the reality of UAP.

Carl Sagan once said about UFOs that "extraordinary claims require extraordinary evidence," and the empirical evidence for the ETH is certainly not proof. However, there is plenty of evidence that warrants reasonable consideration of the possibilities. Let's notice their take on such evidence. See if you agree with Wendt and Duvall.

> "*Physical evidence.* Usually the first objection to the ETH is the lack of direct physical evidence of alien presence. Some ET believers contest this, claiming that the U.S. government is hiding wreckage from a 1947 crash at Roswell, New Mexico, but such claims are based on conspiracy theories that we shall set aside here. Not because they are necessarily wrong (although they cannot be falsified in the present context of UFO secrecy), but because like UFO skepticism they are anthropocentric, only now We know that UFOs are ETs but 'They' (the government) aren't telling. Such an assumption leads critique toward issues of official secrecy and away from the absence of systematic study, which is the real puzzle.
>
> In our view secrecy is a symptom of the UFO taboo, not its heart. While there is no direct physical evidence for the ETH, however, there is considerable indirect physical evidence for it, in the form of UFO anomalies that lack apparent conventional explanations—and for which ETs are therefore one possibility. These anomalies take four forms: ground traces, electro-magnetic interference with aircraft and motor vehicles, photographs and videos, and radar sightings like the Belgian F-16 case. Such anomalies cannot be dismissed simply because they are only indirect evidence for ETs, since science relies heavily on such evidence, as in the recent discovery of extra-solar planets. For if UFO anomalies are not potentially ETs, what else are they?"

Excellent points! We have already discussed these forms of evidence at length in previous chapters, but the point is there is strong evidence worthy of scientific consideration to support the reality of UAP.

> *"Testimonial evidence.* Most UFO reports consist primarily of eyewitness testimony. Although all observation is in a sense testimonial, by itself testimony cannot ground a scientific claim unless it can be replicated independently, which UFO testimony cannot. Such testimony is problematic in other respects as well. It reports seemingly impossible things, much is of poor quality, witnesses may have incentives to lie, honest observers may lack knowledge, and even experts can make mistakes.
>
> In view of these problems skeptics dismiss UFO testimony as meaningless. Problems notwithstanding, this conclusion is unwarranted.
>
> First, testimony should not be dismissed lightly, since none of us can verify for ourselves even a fraction of the knowledge we take for granted. In both law and social science, testimony has considerable epistemic weight in determining the facts. While sometimes wrong, given its importance in society, testimony is rejected only if there are strong reasons to do so. Second, there is a very large volume of UFO testimony, with some events witnessed by literally thousands of people. Third, some of these people were 'expert witnesses'—civilian and military pilots, air traffic controllers, astronauts, astronomers, and other scientists. Finally, some of this testimony is corroborated by physical evidence, as in 'radar/

visual' cases.

> In short, the empirical evidence alone does not warrant rejecting the ETH. It does not warrant acceptance either, but this sets the bar too high. The question today is not 'Are UFOs ETs?' but 'Is there enough evidence they might be to warrant systematic study?' By demanding proof of ETs first, skeptics foreclose the question altogether."

Another line of thinking that prevents a true scientific study of UAP is the inability of skeptics to get past their own doubts and preconceived ideas. Notice these observations:

> "Given the inconclusiveness of the empirical record, UFO skepticism ultimately rests on an a priori theoretical conviction that ET visitation is impossible: 'It can't be true. Therefore, it isn't.' Skeptics offer four main arguments to this effect.
>
> '*We are alone.*' Philosophers have long debated whether life exists beyond Earth, but the debate has lately intensified in response to empirical discoveries like extra-solar planets, water on Mars, and 'extremophile' organisms back home. A thriving discipline of astrobiology has emerged, and the view that life exists elsewhere seems poised to become scientific orthodoxy.
>
> However, this does not mean that (what humans consider) intelligent life exists. The only evidence of that, human beings, proves merely that intelligence like ours is possible, not probable. The Darwinian 'Rare Earth hypothesis' holds that because evolution is a contingent process, human intelligence is a random accident, and the chances

of finding it elsewhere are therefore essentially zero.

This is a serious argument, but there is a serious argument on the other side too, going on within evolutionary theory itself, where the neo-Darwinian orthodoxy is today being challenged by complexity theorists. Rather than contingency and randomness, complexity theory highlights processes of self-organization in Nature which tend toward more complex organisms. If the 'law of increasing complexity' is correct then intelligent life might actually be common in the universe.'

Either way, today it is simply not known. *'They can't get here.'* Even if intelligent life is common, skeptics argue it is too far away to get here. Relativity theory says nothing can travel faster than the speed of light (186,000 miles per second). Lower speeds impose a temporal constraint on ET visitation: at .001 percent of light speed, or 66,960 miles per hour—already far beyond current human capabilities—it would take 4,500 Earth years for ETs to arrive from the nearest star. Higher speeds, in turn, impose a cost and energy constraint: to approximate light speed a spaceship would need to use more energy than is presently consumed in an entire year on Earth.

Physical constraints on inter-stellar travel are often seen as the ultimate reason to reject the ETH, but are they decisive? Computer simulations suggest that even at speeds well below light the colonization wave-fronts of any expanding ET civilizations should have reached Earth long ago. How long ago depends on what assumptions are made, but even pessimistic ones yield ET encounters with Earth within 100 million years, barely a blip in cosmic terms. In short, ETs should be here, which prompts the famous 'Fermi Paradox,' 'Where are They?'

Additionally, there are growing, if still highly speculative, doubts that the speed of light is truly an absolute barrier. Wormholes—themselves predicted by relativity theory—are tunnels through space-time that would immensely shorten the distances between stars. And then there is the possibility of 'warp drive,' or engineering the vacuum around a spaceship, enabling it to skip over space without time dilation. Speculative as these ideas are, their scientific basis is sufficiently sound that research is currently being funded through the 'Breakthrough Propulsion Program' at NASA. They may prove to be wrong or beyond human capacity. But if humans are imagining them just 300 years from our scientific revolution, what might ETs 3,000 years, much less 3,000,000, from theirs be imagining? 'They would land on the White House lawn.' If ETs came all this way to see us, why don't they land on the White House lawn and introduce themselves? After all, if humans encounter intelligent life in our own space exploration, that's what we would do. On this view, the fact that ETs have not is evidence they are not here.

But is it? Again, there is debate. The 'embargo' or 'zoo hypothesis' suggests that ETs might have quarantined Earth as a wildlife preserve. Or, ETs might be interested in contact, but want humans to discover their presence ourselves to avert a violent shock to our civilization. Finally, even humans might not land on the White House lawn. In the popular science fiction show Star Trek, the Federation maintains a policy of 'non-interference' toward lower life forms; might not real space-faring humans adopt a similar policy? Whatever the answer, debates about ET intentions have no scientific basis.

'We would know.' The last skeptical argument is an appeal to human authority: with its panoptic surveillance of the skies the modern state would know by now if ETs were here. Of course, conspiracy theorists think the state does know, but there is no need to embrace this debatable proposition to call the skeptical argument into question. First, skepticism assumes an ability to know the UFO that may be unwarranted. If ETs have the capability to visit Earth, then they may be able to limit knowledge of their presence. Second, no authority has ever actually looked for UFOs, the effect of which on what is seen should not be under-estimated. Finally, in view of pervasive UFO secrecy more is probably known about them than is publicly acknowledged. This does not mean what is known is ET, but it could provide further reason to think so.

Given the stakes, ignoring UFOs only makes sense if human beings can be certain they are not ETs. We have shown there is more than reasonable doubt: the ETH cannot be rejected without significant risk of Type II error. What is actually known about UFOs is that we have no idea what they are, including whether they are alien; far from proving UFO skepticism, science proves its ignorance. With so little science on either side, therefore, the UFO controversy has been essentially theological, pitting ET believers against unbelievers. In this fight, the unbelievers have secured the authority of science, giving them decisive advantage. Their views are taken as fact, while those of believers and agnostics are dismissed as irrational belief.

The UFO taboo is puzzling, we submit, and demands a deeper look at how its 'knowledge' is produced."

Do these arguments sound familiar? Some are, but some are not. This is clearly a thoughtful, deep dive into the psychology of the effect UAP has on human society. It is a challenge to our intellect to be not simply "Believers" but "Knowers." I am all for that! Let's knuckle down and figure this out people!

Clearly, there are things about UAP that we know, and there are things we do not know. There may even be things we can't even imagine and therefore don't even seek to know. Confused? Consider a few of these points:

On the one hand, UFOs appear indeed to be objects, not necessarily in the narrow sense of something hard and tangible, but in the broader sense of natural processes that produce physical effects. The effects are subtle and elusive, which means that UFOs are not unambiguously objects, but radar anomalies and other physical traces suggest something objective is going on.

As an unidentified object the UFO poses a threat of unknowability to science, upon which modern sovereignty depends. Of course, there are many things science does not know, like the cure for cancer, but its authority rests on the assumption that nothing in Nature is in principle unknowable.

UFOs challenge modern science in two ways: (1) they appear random and unsystematic, making them difficult to grasp objectively; and (2) some appear to violate the laws of physics (like the 40g turns in the Belgian F-16 case). This does not mean that UFOs are in fact humanly unknowable, but they might be, and in that respect, they haunt modern sovereignty with the possibility of epistemic failure. To see how this might be uniquely threatening it is useful to compare the UFO to three other cases of what might be seen as unknowability.

One is the **Heisenberg Uncertainty Principle** in quantum theory, which acknowledges inherent limits on the ability to know sub-atomic reality. Since the Uncertainty Principle has not stopped physicists from doing physics, this might seem to undermine our claim that potential unknowability precludes a decision on the UFO as object. Yet, there are known unknowns and unknown unknowns, and here the two cases differ.

Quantum mechanics emerged in a highly structured context of extant theory and established experimental results, and is a systematic body of knowledge that enables physicists to manipulate reality with extraordinary precision.

With quantum theory we know exactly what we cannot know, enabling it to be safely incorporated into modern science. The UFO, in contrast, emerges in a context free of extant theory and empirical research, and raises fundamental questions about the place of human beings in the universe. That we might never know what we cannot know about UFOs makes their potential objectivity more problematic for the modern project.

> "A different problem is presented by *God*, whose existence science also declaims ability to know. Once fiercely contested, the notion that God can be known only through faith not reason is today accepted by religious and secular authorities alike. Since God is not potentially a scientific object, science does not consider the question to be within its purview. Miracles are recognized by the Church, but the criteria by which they are made authoritative are not primarily scientific. UFOs, in contrast, leave unexplained physical traces and as

such fall directly within the purview of modern science."

It is one of the ironies of modern life that it is far more acceptable today to affirm publicly one's belief in God, for whose existence there is no scientific evidence, rather than for UFOs, the existence of which—whatever they might be—is physically documented!

> "It is the triple threat of the UFO that explains states' very different response to it compared to other disruptions of modern norms. By calling into question the very basis of the modern sovereign's capacity to decide its status as exception, the UFO cannot be acknowledged as truly unidentified—which is to say potentially ET—without calling into question modern sovereignty itself.
>
> In this way the UFO exhibits not the standard undecidability that compels a decision, but what might be called a 'meta'-undecidability which precludes it. The UFO is both exceptional and not decidable as exception, and as a result with respect to it the modern sovereign is performatively insecure."

The insecurity is not conscious, but operates at the deeper level of a taboo, in which certain possibilities are unthinkable because of their inherent danger. In this respect UFO skepticism is like denial in psychoanalysis: the government represses the UFO out of fear of what it would reveal about itself.

There is, therefore, nothing for the government to do but turn away, ignore, be ignorant of the UFO, ultimately making no decision at all. Just when needed, most the government is nowhere to be found!

It reminds me of a T-shirt I have from the Mutual UFO

Network (MUFON) that sarcastically states "MUFON-Doing the Air Force's Job Since 1969" (the year project BLUE BOOK was terminated).

So, one of the things that makes me crazy about researching UAP is the whole machine of resistance designed to discredit the entire subject. How has the government been able to pull this off? Is there a cabal of evil doers somewhere in the shadows pulling levers to control the whole world? I think not, but let's see what Wendt and Duvall think on this.

Governmentality and the UFO Taboo

"To this point we have concentrated on the question of 'why?' the UFO taboo, in response to which we have offered a structural answer about the logic of anthropocentric sovereignty. However, there is a separate question of 'how?' the taboo is produced and reproduced, since structural necessity alone does not make it happen. It takes work—not the conscious work of a vast conspiracy seeking to suppress the truth about UFOs, but the work of countless undirected practices that in the modern world make the UFO 'known' as not-ET.

Yet this work too is problematic, because modern governmentality usually proceeds by making objects visible so they can be known and regularized, which in the UFO case would be self-subverting. Thus, what are needed are techniques for making UFOs known without actually trying to find out what they are.

One might distinguish at least four such techniques: (1) authoritative representations, like the U.S. Air Force's claim that UFOs are 'not a national security threat,' the portrayal of ufology as pseudo-science, and the science fictionalization of UFOs in the media; (2) official inquiries, like the 1969 Condon Report, which have the appearance of being scientific but are essentially 'show trials' systematically deformed by a priori rejection of the ETH; (3) official secrecy, which 'removes knowledge' from the system; and finally (4) discipline in the Foucauldian sense, ranging from formal attacks on the 'paranoid style' of UFO believers as a threat to modern rationality, to everyday dismissal of those who express public interest in UFOs, which generates a 'spiral of silence' in which individuals engage in self-censorship instead."

Wow, this blows me away! This is exactly what has been done systematically for more than 70 years. The fact that it has been so effective should be a history lesson for every nation on earth. It is as though they are demonstrating how to control the thinking of an entire free nation!

Let's conclude this intellectual consideration of the political science behind the UAP phenomena with this final thought from Wendt and Duvall.

"The structuralism of our argument might suggest that resistance is futile. However, the structure of the UFO taboo also has aporias and fissures that make it—and the anthropocentric structure of rule that it sustains—potentially unstable.

One is the UFO itself, which in its persistent recurrence generates an ongoing need for its normalization. Modern rule might not recognize the UFO, but in the face of continuing anomalies maintaining such nonrecognition requires work. In that respect the UFO is part of the constitutive, unnormalized outside of modern sovereignty, which can be included in authoritative discourse only through its exclusion."

In other words, those pesky UFOs just don't seem to get the government's hint to scram! Yes, the very fact that UAP continues to present itself to humanity is a factor beyond our control. Clearly, if "they" want it to happen (disclosure of their identity), they will make it so and we can't do a darn thing about it. That is hard to take from the perspective of the world and its nation-state leaders.

Within the structure of modern rule there are also at least two fissures that complicate maintaining UFO ignorance. One is the different knowledge interests of science and the state. While the two are aligned in authoritative UFO discourse, the state is ultimately interested in maintaining a certain regime of truth (particularly in the face of metaphysical insecurity), whereas science recognizes that its truths can only be tentative.

> "Theory may be stubborn, but the presumption in science is that reality has the last word, which creates the possibility of scientific knowledge countering the state's dogma.

> The other fissure is within liberalism, the constitutive core of modern governmentality. Even as it produces normalized subjects who know that 'belief' in UFOs is absurd, liberal governmentality justifies itself as a discourse that produces free-thinking subjects who might doubt it.

> It is in this context that we would place the recent disclosure by the French government

> (and at press time the British too) of its long-secret UFO files (1,600 reports), including its investigations of selected cases, of which the French acknowledge 25 percent as unexplained. Given that secrecy is only a contingent feature of the UFO taboo, and that even the French are still far from seeking systematic knowledge of UFOs, this disclosure is not in itself a serious challenge to our argument. However, the French action does illustrate a potential within liberalism to break with authoritative common sense, even at the risk of exposing the foundations of modern sovereignty to insecurity."

I recommend you obtain a copy of the "COMETA" report that was published in 1999 as a product of a study of French cases that ultimately gave credibility to the ETH as an explanation. Though not an official French government project, it was performed by high ranking military and scientific officials. This is a fascinating document and a must have for your research library.

> "The kind of resistance that can best exploit these fissures might be called militant agnosticism. Resistance must be agnostic because by the realist standards of modernity, regarding the UFO/ET question neither atheism nor belief is epistemically justified; we simply do not know. Concretely, agnosticism means 'seeing' rather than ignoring the UFO, taking it seriously as a truly unidentified object. Since it is precisely such seeing that the UFO taboo forbids, in this context seeing is resistance. The problem is that agnosticism alone does not produce knowledge, and thus reduce the ignorance upon which modern sovereignty depends."

Whether such a science would actually overcome UFO ignorance is unknowable today, but it is only through it that we might move beyond the essentially theological discourse of belief and denial to a truly critical posture.

> "Those who attempt it will have difficulty funding and publishing their work, and their reputations will suffer. UFO resistance might not be futile but it is certainly dangerous, because it is resistance to modern sovereignty itself. But taking UFOs seriously would certainly embody the spirit of self-criticism that infuses liberal governmentality and academia in particular, and it would, thereby, foster critical theory. And indeed, if academics' first responsibility is to tell the truth, then the truth is that after sixty years of modern UFOs, human beings still have no idea what they are, and are not even trying to find out. That should surprise and disturb us all, and cast doubt on the structure of rule that requires and sustains it."

This political science discussion really impacted me personally. I always have asked the question "Why?" It is the very thing that has driven me to become a researcher of the UAP topic. Why is there illogically a denial, even an aversion to seek the answers to the questions that nearly all people ask? Who are we? Why are we here? What does the future hold? Are we alone? Who are these "others"? From where do they come? Do they mean us harm? It is utterly ridiculous, but now more easily understood why this taboo is happening. It is essentially our fear.

You have clearly demonstrated courage by looking into this topic in a serious way. Go for it! I'm glad I did so, and I hope this helps you in your efforts to discover the answers to these profound questions!

> **Note to Self**
> There is no valid
> scientific reason to resist
> the study of UAP.

Chapter 10

Follow the Evidence

Ultimately, I wrote this book for two reasons. The first was to solidify my own findings of this phenomenon and to help guide you through your own search for answers to this most interesting and befuddling topic.

Based on the evidence, the conclusion I have come to so far is this: While there may be no proof, at least available to the public, there is a preponderance of evidence to suggest something very real is happening. I suspect that, if you have not already come to this conclusion yourself, you will as you progress through your personal research.

The UAP enigma is such that there is no certainty of what the big picture may be. We are certain about some details, and we are able to create a likely big picture scenario from those certain details. However, we are left with no choice but to follow the evidence, wherever it leads us.

So, how do we draw any conclusions from this evidence? We must first look at our own bias.

I also must share this bit of personal history because it shapes my point of view and I apply it to my research of UAP. I spent the majority of my life as an avid Bible student. I was a member of a Christian religious organization. I was a leader in that organization, and I learned how to learn, how to think critically, how to write, how to reason with people, and how to speak publicly. These have been very useful skillsets in my life in general, as a husband, father, and even in my career. I am grateful for all of those things. Although, I am no longer an active member of that organization. As a result, my perspective has changed in one respect; that of belief.

Essentially, belief is the acceptance of something as true, without proof. Belief is a necessary function of human intelligence and logic. It is part of the natural process of learning. In religion, we *believe* in the existence of an almighty God. In science, we *believe* in scientific theories, until proven. In business, we often *believe* in management theories. Yes, belief is necessary and is the bridge between gaps in what is thought true and what is later proven true. It moves us along, helps us accomplish, motivates us, gives us emotional support, and many other things.

However, there is a danger with belief. Human beings tend to allow belief to rise to the level of truth when it is not warranted. This can be problematic and even dangerous. Humans can allow belief to be superior to proof. Yes, truth is often denied in the face of strongly help beliefs.

I have heard there are three stages of truth. First, it is ridiculed, then it is violently opposed, and finally it is accepted as self-evident. Is that not true with regard to UAP? This entire book has been about the struggle for the acceptance of UAP as a reality. The truth of the reality of UAP is going through these stages. First, it has been bitterly ridiculed. People pursuing UFO truths even have had their lives ruined in many cases. Secondly, there are cases of violence, (physical, verbal, political) towards UFO researchers, abductees, witnesses, and whistle blowers. And thirdly, more and more people, including scientists, government officials, law enforcement, the military and intelligence community, the public in general, and even the media are beginning to accept UAP as a reality. Make no mistake; UAP is a reality and that is the truth!

So, my viewpoint has changed in this way. I am very careful to **separate belief from proof** in my mind as I research the phenomenon. I really avoid falling into the belief category because I do not want to be wrong. It is too important of a topic to take that risk. In the end, this may be the greatest discovery in the history of mankind. It may even rise to a life and death matter.

There are many people who already believe in the reality of UAP, aliens, and other facets of UAP without proof. As I mentioned earlier, this is a normal part of human intellect and is not necessarily wrong. But there are dangers to placing belief above proof. One can miss really important facts if they are not looking because they think they already know. Although, I don't criticize true believers because they may have more of a reason to believe than I. They may have what they consider proof. Maybe they saw a flying saucer with their own eyes. That would make me a believer! Maybe they have been abducted or contacted, I wasn't there. I don't know, so I can't say they are wrong.

People who are sometimes called "true believers" pejoratively, do invite the ridicule from even fellow Ufologists because they often demonstrate an unreasonable level of gullibility that damages the credibility of not only themselves, but the entire UFO community. As we have discussed, credibility has always been a hard to come by commodity for the UFO community in the view of the public in general. I think that is a valid concern. So, as a member of the UFO research community, you and I have a responsibility to carry our work in an objective and well-balanced way as possible.

So, the most effective way to manage the credibility is to follow the evidence. No one can justifiably refute proven fact. If you have proven fact, then by all means stand up and crow. Tell the world. The truth needs to be exposed, advertised, and put to the forefront. However, when it comes to belief, assumption, theory or other unproven information, proceed more carefully. Openly acknowledge when this is only theory. Follow the evidence and skillfully use it to make a point. Use reason and logic. Place the emphasis on the data. This will serve you well.

Let me share my personal experience as a MUFON investigator to help illustrate the importance of following the evidence. I have been an investigator for about 3 years as of the writing of this book. I have completed the investigation on about 75 cases. The majority of these cases ranged from obvious to less obvious misidentifications, a hoax or two, a couple of abduction cases, to a true unknown or two. So, while I still have much to learn and experience, I have had a diverse array of case evidence to follow.

I should briefly share how I became an investigator for MUFON, because it is something I would highly recommend. Like you, I began to move from merely being a consumer of UFO information motivated by curiosity, entertainment, or serious interest based on a personal experience. Certainly, there is nothing wrong with those motivations, but I began to feel compelled to do more, to engage, participate, even contribute to what I began to consider an important endeavor.

I looked for a way to do so that would match my ability and circumstances. I have been in the civil engineering and geographic information technology fields for decades. My career has been spent in corporate and government organizations. I am a Certified Public Manager. I thrive in large, structured, policy driven, and goal-oriented environments. So, I looked for an organized effort to study UAP as opposed to a self-directed and supported effort. This appealed to me and I thought it was my best chance to succeed.

There are few UAP study organizations that an individual such as I could join. The one that clearly stood out to me was the Mutual UFO Network (MUFON).

MUFON was formed in 1969 and is hailed as the largest UFO related organization in the world. It is an international organization structure similar to a franchise. Although, it is a non-profit organization. It is comprised of a national/international leadership group (i.e., corporate) and then individual chapters (i.e., chapters) that carry out the investigative work. There is criticism of the organization both justified and unjustified, however my experience has been fantastic.

I belong to the Arizona MUFON Chapter in Phoenix, Arizona, USA. In fact, I was recently appointed Arizona MUFON Assistant State Director! The leadership comprised of Jim Mann as State Director, Stacey Wright as Assistant State Director and Dennis Freyermuth as State Chief Investigator has been stellar! Jim recently retired from Arizona State Director, which led to Stacey being appointed State Director and my appointment as

Assistant Director. This chapter even received the first ever "Chapter of the Year award" at the 2017 MUFON Symposium. The leadership, organization, and support given by this team has helped me grow, and given me the opportunity to live out a dream! It is my hope to continue the stellar leadership along with Stacey and Dennis.

The process for becoming an investigator is very reasonable and affordable. You must become a MUFON member, purchase a training manual, take an examination, pass a background check, and finally mentor with an experienced field investigator. Then, case assignments begin. You are never alone though. Each case when submitted as completed is reviewed by the Chief Investigator, who also is an expert in photo analysis. Whether it be him, the Director, Assistant Director, or other field investigators, there is help and support. In addition, there is strong comradery and a social element that has been a pure pleasure. It has even opened other doors for me to be associated with many of the top researchers in the field. Again, I highly recommend this if it sounds like a good fit for you.

In conclusion, I just want to commend you for your earnest interest and efforts to research Unidentified Aerial Phenomena. We need the next generation of researchers to develop. Many of our predecessors are reaching or past retirement age. The work needs to continue, and I hope this book will inspire you to reach out and participate.

We've discussed the identification of UAP, the new modern, less loaded term for UFO. We discussed the need to "calibrate our BS meter" to be able to identify the wheat from the chaff with regards to good and bad UAP information. Remember the five "W's": what, where, when, who, and why?

We also have seen the results of various studies that demonstrate that up to 21% of UAP cases are truly unknown. While, hoaxers create a "boy who cried wolf" effect that has caused many to doubt the reality of UAP. We raised the question when it comes to governments feigned disinterest that "When does a person stop asking a question? Usually when they get an answer . . .

We also noted that there are some credible and serious people, performing credible and serious science, based on credible and serious evidence with respect to UAP. In addition, we identified the need for the highest quality of information, research and reporting to reach the highest quality analysis and conclusions.

We examined several classic cases that revealed that each case has its strengths and weaknesses depending on the quality of the evidence. And that there is no evidence like seeing a UFO or having an encounter to make one a believer!

We also discussed how there is no valid reason that UAP should not be studied. Whether it be science, government, religion, commerce, or any other institution, they are all made up of people. More than anything else, UAP is about people and the impact it can have upon them as individuals or as a society.

And finally, that the evidence is the basis for our beliefs and proofs with respect to UAP.

I hope this beginners guide to researching Unidentified Aerial Phenomena will place you on a direct path to answering the question: UAP; What is it?

Note to Self
It's really about the evidence!

Sources

1. Randles, Jenny. UFO Study: a Handbook for Enthusiasts. London: Robert Hale, 1981.

2. "National Aviation Reporting Center on Anomalous Phenomena." National Aviation Reporting Center on Anomalous Phenomena. Accessed January 30, 2020. https://www.narcap.org/.

3. "UFODATA Project." UFODATA Project. Accessed January 30, 2020. http://ufodata.net/.

4. Carey, Benedict. "The Fame Motive." The New York Times. The New York Times, August 22, 2006. https://www.nytimes.com/2006/08/22/health/psychology/22fame.html.

5. Friedman, Stanton. Flying Saucers and Science. New Page Books, 2008.

6. Phillips, Ted. "Physical Traces Associated with UFO Sightings." Edited by Mimi Hynek. Center for UFO Studies. Accessed January 30, 2020. http://www.cufos.org/.

7. "Roswell UFO Retrieval - 1947." MUFON. Accessed January 30, 2020. https://www.mufon.com/roswell-ufo-retrieval---1947.html.

8. Emenegger, Robert. UFOs, Past, Present, and Future. New York: Ballantine Books, 1974.

9. Walton, Travis. Fire in the Sky: The Walton Experience. Snowflake, AZ: Skyfire Productions, 2010.

10. James Fox, Out of the Blue, Springdale, AR, Hanover House, 2003

11. Alexander Wendt and Raymond Duvall, Sovereignty and the UFO, Thousand Oaks, CA, Sage Publications, 2008

www.ingramcontent.com/pod-product-compliance
Lightning Source LLC
Chambersburg PA
CBHW071413210526
45465CB00001B/368